Cows in the Maze

Ian Stewart is Emeritus Professor of Mathematics at Warwick University. An active research mathematician, he is also a world renowned popular writer of mathematics. His many books include *Professor Stewart's Cabinet of Mathematical Curiosities, How to Cut a Cake, From Here to Infinity, What Shape Is a Snowflake?* and *The Science of Discworld* and *The Science of Discworld II* (both with Terry Pratchett and Jack Cohen).

Cows in
the Maze

IAN STEWART

OXFORD
UNIVERSITY PRESS

OXFORD
UNIVERSITY PRESS

Great Clarendon Street, Oxford OX2 6DP

Oxford University Press is a department of the University of Oxford.
It furthers the University's objective of excellence in research, scholarship,
and education by publishing worldwide in

Oxford   New York

Auckland Cape Town Dar es Salaam Hong Kong Karachi
Kuala Lumpur Madrid Melbourne Mexico City Nairobi
New Delhi Shanghai Taipei Toronto

With offices in

Argentina Austria Brazil Chile Czech Republic France Greece
Guatemala Hungary Italy Japan Poland Portugal Singapore
South Korea Switzerland Thailand Turkey Ukraine Vietnam

Oxford is a registered trade mark of Oxford University Press
in the UK and in certain other countries

Published in the United States
by Oxford University Press Inc., New York

British Library Cataloguing in Publication Data

Data available

Library of Congress Cataloging in Publication Data

Data available

Typeset by SPI Publisher Services, Pondicherry, India
Printed in Great Britain
on acid-free paper by
Clays Ltd., St Ives Plc

ISBN 978-0-19-956207-7

1 3 5 7 9 10 8 6 4 2

Contents

Introduction

The cows are back.

If you're new to this game, or haven't been paying attention, *Cows in the Maze* is Oxford University Press's third collection of my Mathematical Recreations columns from *Scientific American* and its French edition *Pour La Science*. The French edition typically contains its own special material, and for a time I wrote six columns a year for the American edition and another six for the French. And there are two earlier collections from other publishers.

Oh yes, those cows.

When we were putting together Oxford University Press's first collection, *Math Hysteria*, the editors decided to make the book seem even more friendly by providing cartoons for each chapter, and of course the cover. In a stroke of genius, they decided to ask Spike Gerrell. One of the chapters was on 'counting the cattle of the Sun', a fiendishly complicated puzzle whose answer has 206,545 digits and was first discovered in 1880. There are reasons to believe that perhaps Archimedes had not intended it to be *that* fiendish...but you can never tell with Archimedes.

Anyway, Spike seized upon this hint of a cow-y theme, because he does particularly comely cows. On the cover, one was jumping over the Moon, and three were wearing blindfolds – well, hoods, actually. If you look at the book's spine you will see one cow peeping round the corner at you.

The next collection, *How to Cut a Cake*, was a cow-free zone, though Spike did come up with some chessboard horses, an entangled cat – in a phone cord, not related to Schrödinger or anything quantum – and a bemused rabbit. The opportunity to compensate the cows for this injustice presented itself when we decided to put together another collection, and one of the possible topics was *Cows in the Maze*. Saved us the trouble of thinking of a title, too.

Now, you may have thought that mathematics is a pretty serious business, and a herd of cows rampaging through a maze, watched by a gang of engineers who are either building the maze or demolishing it, lacks the proper *gravitas*. But, as I've said many times now, 'serious' need not equate to 'solemn'. Mathematics is indeed a serious business: our civilization could not possibly function without it – an aspect of the subject that admittedly is news to many, but easy enough to prove to anyone who wants to know. For that reason, mathematics is so serious that we all need to chill out a bit, and stop getting so uptight about decimal points and fractions and parallelograms (do they do those nowadays?) that we conceal the great secret that makes the whole subject much more palatable.

Namely: it's fun.

Even the serious stuff is fun, in a serious kind of way. Hardly anything can beat that amazing feeling when the little light bulb in your head goes off and you suddenly *understand* what makes a piece of mathematics tick. Mathematical research – a big part of my job when I'm not writing books – consists of 99% banging your head against a meta-phorical brick wall, and 1% suddenly realizing why it's all totally obvious and you've been extraordinarily stupid. *Flash!* goes the light bulb, and you shrug off the feeling of foolish-ness on the grounds that 99.99% of the human race wouldn't understand the problem, let alone the answer, and mathem-atics always looks easy once you've understood it.

One of the reasons I became a mathematician was the monthly mathematics column in *Scientific American* – then titled 'Mathematical Games' and written by the inimitable Martin Gardner. Gardner wasn't a mathematician, but it would be too limiting to call him a journalist. He's a writer, whose interests include puzzles, magic (of the stage variety), philosophy, and exposing the idiocies of pseudo-science. His Mathematical Games column worked precisely because he *wasn't* a mathematician, but he had an uncanny instinct for the interesting, the curious, and the significant. He is an impossible act to follow, and I've never tried to do that. But it was Gardner who showed me that mathematics is much broader and richer than anything I'd been exposed to at school.

I'm not complaining about school maths. I had a series of excellent teachers, one of whom – his name was Gordon Radford – used up most of his spare time teaching me and a few friends the same lesson that I was getting from Gardner: there's a lot more to maths than the textbooks lead you to assume. School gave me the technique, but Gardner gave me the *passion*. In her autobiography, *To Talk of Many Things*, Dame Kathleen Ollerenshaw – one of Britain's truly great mathematics educators – recounts an incident when she was at school, and let slip her hope of discovering some new mathematics. One of her fellow students expressed a contrary opinion: why bother, when there was already too much of it? I side with Dame Kathleen. In fact, one chapter shows that her ambition was fulfilled, even though her career track went into education and local government. She was 82 years old at the time, and that was ten years ago.

Cows in the Maze can be read in any order: each chapter stands alone, and you can skip anything that bothers you. (Here's another great mathematical secret, which I was fortunate to learn at an early age: don't get hung up on difficult details, plough ahead anyway. Often light then dawns, and if not, you can always go *back* and try again.) The only exception is a series of three chapters (originally two columns, but one was gigantic so I split it) on the mathematics of time travel.

The topics are diverse – it's not a textbook, it's a celebration of the joy of mathematical investigation and discovery. Some chapters are in 'story' format, others are straight descriptions. I had to stop presenting the column in story format when my space in the American magazine was cut from three pages to two. The French continued to indulge my sense of narrative, every alternate month when there wasn't an American column, until the Americans let me write a column every month. And, cows notwithstanding, the discerning reader will find a great diversity of genuine mathematics scattered through these pages: number theory, geometry, topology, probability, combinatorics ... and several areas of applied mathematics, including fluid mechanics, mathematical physics, and animal locomotion.

The columns benefited from a lively correspondence with readers, and by the end they were providing about half of the ideas for topics. We started a 'Feedback' section, and I've included readers' suggestions in most chapters. I've tried to preserve the feel of the originals, while bringing them up to date and removing any errors or ambiguities that I know about. I've also introduced a new feature to reflect the increasing influence of the Internet: references to interesting websites.

I am grateful to my editor Latha Menon and everyone else at OUP who let themselves be persuaded to sanction my further romps with Spike's cows, to Spike for a cow-bedecked cover, to Philippe Boulanger who started it all by letting me loose between the covers of *Pour La Science*, and to *Scientific American* for helping me to fulfil a childhood dream.

Coventry, September 2009   Ian Stewart

Figure Acknowledgements

46 Reproduced with permission. © Andrew David-hazy.

64 Reproduced with permission. © Science Museum/SSPL.

69 Reproduced with permission. © Nature and Jonathan Callan.

73 Reproduced with permission. © Sloan Digital Sky Survey.

74 Reproduced with permission. © NASA.

91–92 Reproduced with permission. © Dr. Schaffer and Mr. Stern Dance Ensemble.

1

The Lore and Lure of Dice

Dice...They seem such simple things, just cubes with numbers on them. The ancients used them for gambling, and for divining the will of the gods. The mathematics of dice is more recent, part of a wider understanding that chance has its own patterns. If you know how to look for them.

THE DIE, more commonly known by its plural 'dice', is one of the earliest known gambling aids. The Roman historian Herodotus claimed that dice were introduced by the Lydians in the time of King Atys, but Sophocles disagreed, crediting their invention to a Greek called Palamedes, allegedly during the siege of Troy. It may seem plausible that dice were invented to give the bored besiegers something to do while they waited for the Trojans to surrender, but the credit must go to others. Dice have been found in Chinese remains from about 600 BC. Archaeologists have discovered cubical dice, to all intents and purposes just like today's, in Egyptian tombs dating from 2000 BC. Other finds go back to 6000 BC. Dice seem to be one of those basic forms that originated independently in many different cultures. The cubical shape, however, is not unique. Dice of many shapes and with many strange markings have been used by North American Indians, South American cultures such as the Aztecs and Mayas, Polynesians, Inuits, and many African tribes. They have been made from materials ranging from beaver teeth to porcelain.

The game of dungeons and dragons uses dice shaped like regular solids.

Dice are such simple things, but their possibilities are almost endless.

To stop this chapter taking over the whole book, I'm going to focus exclusively on standard, modern dice. These are, of course, cubical in shape, and usually have rounded edges and corners. Their key feature is a pattern of spots on each face, the numbers of spots being 1, 2, 3, 4, 5, and 6. Spots on opposite faces sum to 7, so the faces come in three pairs: 1 and 6, 2 and 5, 3 and 4. Up to rotations of the cube, there are exactly two possible arrangements with this property (Figure 1), and one is the mirror image of the other. Nowadays virtually all dice of western manufacture are like Figure 1a, in which the faces 1, 2, 3 cycle round their common vertex in the anti-clockwise direction. I am told that in Japan, dice with this handedness are used in all games except Mah-Jong, where mirror-image dice of Figure 1b are used instead. Oriental dice have a much larger spot for the number 1, and some spots may be red instead of black, depending on the culture.

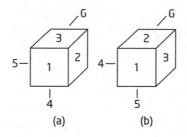

FIG 1 The two different ways to number dice.

Dice are often thrown in pairs, and a fundamental fact here is the probability of getting a given total. To calculate these probabilities – on the assumption that the dice are 'fair', meaning that each face has a probability 1/6 of coming up on top – we work out how many ways there are to achieve a given total. Then we divide that by 36, the total number of pairs, taking into account which die is which. To do this it helps to imagine that one die is red and the other blue. Then a total of 12, say, can occur in only one way: red die = 6, blue die = 6. The probability of a total of 12 is therefore 1/36. A total of 11, on the other hand, can occur in two ways: red die = 6, blue die = 5, *or* red die = 5, blue die = 6. Its probability is therefore 2/36 = 1/18.

This may seem obvious, but dice are usually indistinguishable, and colouring them is a bit artificial. As illustrious a thinker as the great mathematician and philosopher Gottfried Leibniz thought that the probabilities of throwing 11 and 12 must be the same. He argued that there is only *one* way to throw 11: one die = 6, the other = 5. There are several problems with this line of attack, however. Perhaps the most significant is that it disagrees wildly with experiment, in which 11 comes up about twice as often as 12. Another is that it leads to the unlikely conclusion that the probability that two dice throw *some* total (whatever it may be) is less than one. Or, if you don't like that interpretation, it implies that the probability of throwing 12 is bigger than 1/36.

Figure 2 shows the probabilities for all totals from 2 to 12. One game in which an intuitive feel for these probabilities is crucial is craps, which dates from the 1890s. Here one player, the shooter, puts up a sum of money. The others 'fade' it – that

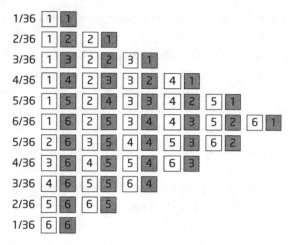

FIG 2 The probabilities of totals for two dice.

is, they bet an amount of their own choice. If the total faded is less than the shooter's initial bet, then the shooter reduces the bet to match that total. The shooter then rolls the dice. A score of 7 or 11 (natural) on the first roll wins outright; a score of 2 (snake eyes), 3, or 12 (craps) loses. Otherwise the shooter's initial score, one of the numbers 4, 5, 6, 8, 9, 10, becomes his 'point'. He continues to roll, aiming to score the point again before he throws 7 (craps out). If he succeeds, he wins all the money; if he fails, he loses.

From Figure 2 and a few other considerations it can be calculated that the shooter's chance of winning is 244/495, roughly 49.3%. This is *just* less than evens (50%). Professional gamblers can turn this slight disadvantage into an advantage by two methods. One is to accept or reject various 'side-bets' with other players, exploiting superior knowledge of the odds. The other is to cheat, using sleight of hand to introduce rigged dice into the game.

Dice can be rigged in several ways. Their faces may be subtly shaved so that their corners are not right angles, or they can be 'loaded' with weights. Both of these techniques make some throws more probable than others. More drastically, the standard dice may be replaced by 'tops': rigged dice that come in several varieties. For example, the die may bear only three distinct numbers of spots, with opposite faces having identical numbers. Figure 3 shows an example with the faces 1, 3, 5 only. Because each player sees at most three faces of a die at any given instant, and because no two adjacent faces of tops have the same number of spots, nothing appears amiss to a cursory glance. However, it is not possible to ensure that the arrangements at all vertices cycle in the 'correct' order. Indeed, if the order is 135 anticlockwise around one vertex, then it must be 135 clockwise around an adjacent vertex, as Figure 3 shows. So an alert player can detect the subterfuge.

Tops can be used in craps for several purposes. A pair of 135s, for instance, can never throw 7, so with these a player can

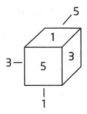

FIG 3 'Tops' – how to cheat.

never crap out. A combination of one 135 and one 246 cannot produce an even total, so with these dice, a player cannot make a point of 4, 6, 8, or 10. Tops must be used sparingly if their presence is to be undetected – even the most naive of players will eventually start to wonder why they keep throwing odd totals. So the rigged dice are usually switched rapidly in and out, to change the odds just a little in the favoured direction. There are also 'one-way tops' in which only one number of spots occurs twice. Instant recognition of the arrangement of the spots on a die is essential knowledge for professional gamblers, because it can help them detect tops.

Many conjuring or party tricks use dice. A lot of them are based on the rule that opposite faces sum to 7. Martin Gardner describes one of them in his *Mathematical Magic Show*. The magician turns her back and asks a member of the audience to roll three standard dice and add up the top faces. Then the victim is told to pick up any die and add its bottom number to the total. Finally, the victim rolls the same die again and adds its top number to the previous total. Now the magician turns round and immediately states what the result was – even though she has no idea which die was chosen.

How does this work? Suppose that the dice have totals a, b, and c, and that (say) die a is chosen. The initial total is $a + b + c$. To this is added $7 - a$, making $b + c + 7$. Then a is thrown again, giving d, and the final result is $d + b + c + 7$. The magician then looks at the three dice, which total $d + b + c$ – so all she has to do is quickly add them up, and add 7.

Henry Ernest Dudeney, the great English puzzlist, includes a trick of a different kind in his book *Amusements in Mathematics*. Again the magician asks for three dice to be thrown while her back is turned. This time the victim is asked to double the value of the first die and add 5; then multiply the result by 5 and add the value of the second die; then multiply the result by 10 and add the value of the third die. Upon being told the result, the magician immediately says what the three dice were. The result, of course, is now $10(5(2a + 5) + b) + c$, or $100a + 10b + c + 250$. So the magician subtracts 250 from the result, and the three digits of the answer are the numbers on the dice.

Games with dice need not involve any random element. One such game begins by one player choosing a 'target' number, such as 40. The other player places a single die on the table, with some chosen face on top – say 3. This value starts a running total. The other player may now roll the die through a quarter turn – which here reveals either 1, 2, 5, or 6. Whatever comes up top is added to the running total. If, for instance, the second player turns the die to show 2, then the running total becomes $3 + 2 = 5$. The players take turns to roll the die through a quarter turn, in whatever direction they wish, and the running total accumulates. The first player to make the running total bigger than the target loses.

There is a systematic method for analysing such games, explained in detail in my book *Another Fine Math You've Got Me Into*. The idea is to divide positions of the game into two classes, 'win' and 'lose', and work backwards from the end, using the following two principles:

- If *any* move from the current position leads to a winning position (for the other player) then the current position is a losing one.

- If *some* move from the current position leads to a losing position (for the other player) then the current position is a winning one.

For instance, if the current running total is 39 and face 1 is uppermost, then the next player has no choice but to exceed 40, so this position is a winning one. In order actually to *win*, you have to play the appropriate move.

In carrying out this calculation, it is best to work with the difference between the current total and the target – that is, the 'effective target' from that stage onwards. In the above example, the effective target is $40 - 39 = 1$, and whatever move the next player makes, they must exceed it. On the other hand, if face 2 is uppermost when the effective target is 1, then the next player can turn the die so that 1 is on top, and win.

The table below summarizes the status of various states of the game, for effective targets between 0 and 25. Here the state – the face that is uppermost – is shown at the left of the rows, the effective total is at the top of the columns, and each column either shows 'L' for a losing position, or a list of winning moves for a winning position. Notice that states 1 and 6 are in effect the same, since they lead to the same four possible moves 2, 3, 4, 5. The same goes for states 2/5 and 3/4. So the table has only three rows.

Effective target status:	1	2	3	4	5	6	7
1 or 6	L	2	3	4	5	3	234
2 or 5	1	1	3	4	L	36	346
3 or 4	1	12	L	L	5	6	26

Effective target status:	8	9	10	11	12	13	14	15	16
1 or 6	4	L	5	23	34	4	5	3	234
2 or 5	4	L	1	3	34	4	L	36	34
3 or 4	L	L	15	2	L	L	5	6	2

Effective target status:	17	18	19	20	21	22	23	24	25
1 or 6	4	L	5	23	34	4	5	3	234
2 or 5	4	L	1	3	34	4	L	36	34
3 or 4	L	L	15	2	L	L	5	6	2

I've laid out the tables to emphasize the main feature: columns 17–25 are the same as columns 8–16. This pattern, once it becomes established, must repeat indefinitely, so columns 26–34, 35–43, 44–52, and so on, are also the same as 8–16. The reason is that any move reduces the effective target by 6 at most, so the entries in a given column depend only on those in the six columns to its left. So as soon as a block of six (or more) consecutive columns repeats entries seen in a previous block, the pattern must repeat indefinitely.

Such repetitions are to be expected in all games of this general kind, because there are only finitely many possible columns. But we're lucky that the repeating block occurs so soon, and is so short. It leads to a complete, but far from intuitive, prescription for a winning strategy. Take your chosen target and repeatedly subtract 9 until you first get into the range 0–16. Then look in the resulting column to see whether the position is a win or lose – and if it's a win, play one of the recommended winning moves.

For instance, suppose the target is 1000. Subtracting 9s repeatedly we get down to 19, which is still bigger than 16, and finally to 10, where we stop. Column 10 tells us that we can always make a winning move. If the state is 1/6 then we move the die to show 5; if the state is 2/5 we move it to 1; and if the state is 3/4 we move it to 1 or 5. Keep repeating this procedure, and eventually you must win.

If you're unlucky, and the initial position is a losing one, you have to hope that your opponent doesn't know the strategy. Make any move you like, wait till they've made theirs, and repeat the calculation. You should soon hit a winning position, unless a miracle is in progress, after which you control the game completely. With a moderately heroic effort, you can commit the entire table to memory. Or you can simplify it by remembering only one winning move for each state, rather than the whole list. In fact, if you do that intelligently, you can ignore all columns after the eleventh, reducing the amount to be learned to something fairly manageable.

Other dice problems involve modified dice with non-standard numbering. For example: can you think of a way to label two dice, using only numbers 0, 1, 2, 3, 4, 5, or 6, to get a pair of dice such that all totals from 1 to 12 are equally likely? (See the end of the chapter for the answer.) Perhaps the most counter-intuitive dice phenomenon is that of 'non-transitive dice'. Make three dice A, B, and C, numbered like this:

A: 3 3 4 4 8 8
B: 1 1 5 5 9 9
C: 2 2 6 6 7 7

Then, in the long run, B beats A. In fact, die B throws a higher total than A with probability 5/9. Similarly C beats B with probability 5/9. So obviously C beats A, right? No, A beats C with probability 5/9. The next table justifies these assertions: it lists the winner for each combination of dice. For example if B is playing C, look at the second array in the table. Suppose B throws 5 and C throws 6. Then C has the higher throw, so C wins. Therefore column 5, row 6 of the array is C.

	A	3	4	8
B				
1		A	A	A
5		B	B	A
9		B	B	B

	B	1	5	9
C				
2		C	B	B
6		C	C	B
7		C	C	B

	C	2	6	7
A				
3		A	C	C
4		A	C	C
8		A	A	A

In the first array there are 5 Bs and 4 As, so B beats A with probability 5/9, as I claimed. In the second array there are 5 Cs and 4 Bs, so C beats B with probability 5/9. In the third array there are 5 As and 4 Cs, so A beats C with probability 5/9.

You can make a fortune with a set of such dice! Let your opponent choose one; then you choose whichever one beats it (in the long run, with probability greater than evens). Repeat. You will win on 55.55% of all plays. Yet your opponent has a free choice of the 'best' die!

A word of warning, though: don't place *too* much reliance on probability theory without making the rules of the game *very* precise. In his marvellous little book *The Broken Dice*, Ivar Ekeland tells the story of two Nordic kings who played dice to decide the fate of a disputed island. The King of Sweden rolled two dice and scored a double 6. This, he boasted, was unbeatable, so King Olaf of Norway might as well give up.

Olaf muttered something to the effect that he, too, might score a double 6, and cast his two dice. One turned up 6; the other split into two pieces, one showing a 1 and the other a 6. Total: 13! All of which goes to show that what you think is possible depends upon how you model the problem.

If the tale is true, King Olaf was extraordinarily fortunate. A few cynics think that Olaf rigged the whole scam.

FEEDBACK

Many readers wrote in with their own variations on the set of three 'non-transitive' dice in the November 1997 column. My dice had faces (each occurs twice) as follows: A:(3,4,8); B:(1,5,9); C:(2,6,7). Then B beats A with probability 5/9, C beats B with probability 5/9, and A beats C with probability 5/9. George Trepal of Gehring, Florida pointed out that these sets of numbers, suitably arranged, form the columns of a magic square - an array of numbers whose rows, columns, and diagonals all add up to the same amount. The magic square concerned is

$$
\begin{array}{ccc}
8 & 1 & 6 \\
3 & 5 & 7 \\
4 & 9 & 2
\end{array}
$$

Moreover, there is a curious 'duality': if the rows of this square are used for faces on dice instead, say A:(8,1,6); B:(3,5,7); C:(4,9,2) – again with each face occurring twice if you want six-sided dice instead of unorthodox three-sided ones – the resulting set is again non-transitive, and A beats B with probability 5/9, B beats C with probability 5/9, and C beats A with probability 5/9.

With the magic square

$$
\begin{array}{ccc}
8 & 1 & 9 \\
7 & 6 & 5 \\
3 & 11 & 4
\end{array}
$$

the results are interestingly different. For the rows, A beats B with probability 6/9, B beats C with probability 6/9, and C beats A with probability 5/9. For the columns, B beats A with probability 5/9, C beats B with probability 5/9, and A beats C with probability 5/9.

Trepal's best set – using the smallest numbers – follows the 6/9, 6/9, 5/9 pattern, and is: A:(1,4,4); B:(3,3,3); C:(2,2,5). Zalman Usiskin of the University of Chicago raised and answered a natural question. Can you make the advantages bigger than 5/9? More precisely, given three non-transitive six-sided *loaded* dice, what is the largest possible probability p for which all three pairs provide a win with probability at least p? By 'loaded' I mean that the faces need not appear with equal probability. The answer is a new occurrence of a famous number, the golden number

$$\phi = \frac{1+\sqrt{5}}{2}.$$

Suppose that:

A scores 4 with probability $\phi - 1$ and 1 with probability $2 - \phi$;

B always scores 3;

C scores 2 with probability $\phi - 1$ and 5 with probability $2 - \phi$.

Then A beats B, B beats C, and C beats A, all with probability $\phi - 1$, which is approximately 0.618. This is significantly larger than $5/0.9 = 0.555$, and it is the largest advantage possible.

Loaded dice can be simulated, to high accuracy, by fair dice with lots of faces, by repeating each number suitably many times. Using an icosahedron, with 20 faces, we can achieve a figure of $16/25 = 0.64$, as follows:

A has 4 on 12 faces and 1 on 8 faces;

B has 3 on all 20 faces;

C has 2 on 12 faces and 5 on 8 faces.

ANSWER

To make two dice for which all totals from 1 to 12 equally likely, one must have faces 1, 2, 3, 4, 5, 6, and the other 0, 0, 0, 6, 6, 6.

WEBSITES

GENERAL:

  http://en.wikipedia.org/wiki/Dice
  http://mathworld.wolfram.com/Dice.html

NON-TRANSITIVE DICE:

  http://en.wikipedia.org/wiki/Nontransitive_dice

HISTORY:

  http://hometown.aol.com/dicetalk/polymor2.htm

RIGGED DICE:

  http://homepage.ntlworld.com/dice-play/

2

Pursuing Polygonal Privacy

Some of the most difficult questions in mathematics are inspired by everyday life. Who would have thought that the simple act of building fences can suggest problems that no one has yet been able to solve?

ONE OF THE most appealing areas of mathematics, full of simple problems whose solutions are currently unknown, is combinatorial geometry. In such problems, the aim is to find arrangements of lines, curves, or other geometric shapes that achieve some objective in the most efficient manner possible. For example, the Mother Worm's Blanket problem[1] asks: what is the shape of the smallest area that can cover a curve of unit length, no matter how that curve is arranged? Although many candidate shapes have been proposed, no such shape has yet been proved to have minimal area, and it remains possible that the problem has no solution at all. Recreational mathematicians can have a lot of fun with such questions, because there is plenty of scope for experiment and ingenuity. Even if it is not possible to prove that some particular shape is the best possible, you can often find improvements on those that were known previously.

This chapter concentrates on a puzzle known as the Opaque Square Problem, along with several fascinating

---

[1] See *Game, Set and Math*, Chapter 1.

variations. It was brought to my attention by Bernd Kawohl (Cologne), and the discussion is based on an article he sent me. Suppose you own a square plot of land, whose sides, for simplicity, are assumed to have unit length. For some reason best known to you – privacy, say – you want to build a fence on your land that will block any straight line of sight passing across it. Moreover, to save money, you want the fence to be as short as possible, subject to blocking every line of sight. How do you arrange the fence?

The fence can be as complicated as you like, with lots of different pieces, joined together however you wish. The pieces of fence can be curved or straight. In fact, it could be any shape for which some generalization of the concept 'length' makes sense.

Perhaps the most obvious solution is to build a fence round the entire perimeter, which gives a total length of 4 (Figure 4a). A few moments' thought reveals an improvement: leave out one side to create a square-cornered U shape (Figure 4b). Now the length reduces to 3. This is in fact the shortest fence if we make the additional assumption that the fence must be a *single* polygonal or curved line. Why? Because every fence that renders the square opaque must contain all four corner points (otherwise there is a 'line of sight' passing through a

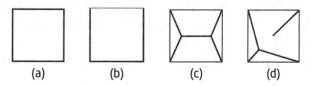

(a)   (b)   (c)   (d)

FIG 4 Opaque fences for the square.

corner) and the shortest single curve that contains all four corners is composed of three sides of the square.

However, a more complicated fence exists with length $1 + \sqrt{3} = 2.732$, as in Figure 4c. The angles between the lines here are all 120°. Arrangements of this kind, in which the fence is connected, are called Steiner trees, and it has long been known that 120° angles keep the length of the tree as short as possible.[2] This is the shortest *connected* fence. Nonetheless, we haven't finished. If we allow the fence to have several disconnected pieces, the total length can be reduced to 2.639 as in Figure 4d. Here the three lines in the upper half of the diagram again meet at angles of 120°. This final attempt is widely believed to be the shortest possible opaque fence, but nobody has yet found a proof.

Indeed, it has not even been proved that a shortest opaque fence exists. The main problem in proving existence is that it might (perhaps!) be possible to keep shortening the length by making the fence more and more complicated. Vance Faber and Jan Mycielski have proved that for any given number of connected components, there exists a shortest opaque fence. What is not known is whether the minimal length keeps shrinking as the number of components increases without limit, or whether a fence with an infinite number of components can out-perform all fences with finitely many components. It seems unlikely that either of these things can happen, but neither has yet been ruled out.

Kawohl has given a lovely proof that Figure 4d is the shortest fence having exactly two components. First, he

[2] See *How to Cut a Cake*, Chapter 12.

shows that one component must contain three corners of the square and the other must contain the remaining corner. The first component must therefore be the shortest Steiner tree linking three corners, and it is known that this has the shape shown in the upper part of the figure. The convex hull of this shape – the smallest convex region that contains it – is the triangle formed by cutting the square in two along a diagonal. The second component must be the shortest curve that joins the fourth corner to this triangle, and this is clearly the diagonal line from that corner to the centre of the square.

What about shapes other than the square? If the plot of land is an equilateral triangle, then the shortest opaque fence is a Steiner tree, formed by joining each corner to the centre along a straight line (Figure 5a). If the plot is a regular pentagon, then the shortest known opaque fence comes in three pieces, as in Figure 5b. One piece is a Steiner tree linking three adjacent corners of the pentagon. The second is a straight line joining the fourth side to the convex hull of the first three corners. The third is a straight line joining the final

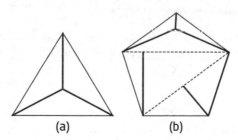

(a)   (b)

FIG 5 Opaque fences for the equilateral triangle and the regular pentagon and hexagon.

corner to the convex hull of the first four. Again, no proof exists that this fence has minimal length, but no shorter opaque fence has been found.

For the regular hexagon, the best fence known is similar, but because the corner angles of the hexagon are 120°, the Steiner tree becomes a series of edges of the hexagon. In fact, it consists of three consecutive edges, linking four adjacent corners together. Then the second component of the fence is the shortest line joining a fifth corner to the convex hull of the first four, and the third component is the shortest line joining the sixth corner to the convex hull of the first three.

It has not been *proved* that this fence is optimal, but the construction extends to give a conjectured minimal fence for any regular polygon with an even number of sides (Figure 6). Divide the polygon into two by a diameter joining two opposite corners. The first component of the fence is formed from all of the edges that lie in that half, forming the polygonal analogue of a semicircle. The second component is the shortest line linking the next corner to the convex hull of the

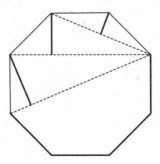

FIG 6 Conjectured shortest opaque fence for an even-sided regular polygon.

first component, the third component is the shortest line linking the next corner to the convex hull of the first two components, and so on.

A polygon with a large number of sides is very close to a circle, and we can ask for the shortest fence that makes a circle opaque. By choice of units, we may assume that the circle has unit radius. The simplest fence that comes to mind is the circumference of the circle, of length $2\pi = 6.283$ (Figure 7a). However, if the fence is permitted to lie *outside* the plot of land, we can do better. Remove half the circumference to leave a semicircle, of length π, and extend it by adding two lines of length 1 that are tangent to the circle at the ends of the semicircle, forming a U (Figure 7b). This is an opaque fence for the circle, and its length is $\pi + 2 = 5.142$.

It can be proved that Figure 7b is the shortest possible fence if we insist that the fence be a single curve – no branch points and all in one piece. There is another way to state its 'opaqueness' property.[3] Suppose that a straight pipe or telephone line

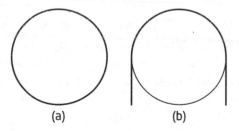

(a)   (b)

FIG 7 Opaque fences for the circle. (a) The circle itself, length 2π for a circle of radius 1. (b) A shorter fence, length $\pi + 2$.

[3] See *Math Hysteria*, Chapter 6.

is known to pass within distance 1 of some specific point: what is the shortest trench we can dig that is guaranteed to find it? We know that the pipe must cross the circle of unit radius centred at that point, and must therefore hit any opaque fence for that circle. So we should dig a trench in the form of an opaque fence.

With the trench version of the puzzle it is natural to allow the trench to go outside the circle – but fences are normally built on the owner's land, not on their neighbours'. Kawohl shows that the shortest opaque fence lying entirely inside the circle of unit radius also has length no greater than $\pi + 2$. He does this by considering the conjectured fence for an even-sided polygon with a large number of sides, closely approximating the unit circle. A trigonometric calculation proves that the length of a fence like the one shown in Figure 6, but with more sides to the polygon, is then very close to $\pi + 2$. The difference can be made as small as we please by taking a large enough number of sides.

There is much here for the amateur to investigate. Are the conjectured fences truly the shortest possible, or is there a way to shorten them further? Can anything be proved about the conjectured solutions? What about other shapes – arbitrary polygons (convex or not), ellipses, semicircles…And what about the same problem in three dimensions: the opaque cube and sphere? Now the aim is to minimize the total area of the fence.

FEEDBACK

Martin Gardner raised the problems of the opaque cube and sphere in 1990, and Kenneth A. Brakke of Susquehanna University tackled them in 1992 (see Further Reading and Website). Brakke's best solution for a unit cube has an area of 4.2324.

WEBSITE

OPAQUE CUBE:

http://www.susqu.edu/brakke/opaque/default.html

3

Making Winning Connections

Some mathematical games are *truly* mathematical, and there is no better example than Hex. All you have to do is place your counters on a honeycomb-patterned board, and connect two opposite edges. Easy? There's an entire book devoted to it.

WHAT DO A Danish poet-mathematician and a Nobel laureate have in common? One of the best mathematical board games ever invented, that's what. Nowadays it's usually called Hex, but its early incarnations bore a variety of names. Cameron Browne's *Hex Strategy* takes a comprehensive look at Hex and how to win it. Hex is at least as addictive as front-line computer games, and gives your brain a far more stimulating workout.

Hex is a two-player game, played on a board made from hexagonal cells, arranged in the shape of a rhombus (Figure 8). The standard board size is 11 × 11, but other sizes provide entirely playable games. Each player 'owns' two opposite edges of the board; the four corner cells are joint property. One player has a stock of black counters, the other a stock of white counters – stones from the oriental board game Go are ideal.

The rules are astonishingly simple. Players take turns to place one of their counters on an unoccupied cell of the board – who starts is decided by tossing a coin or any other

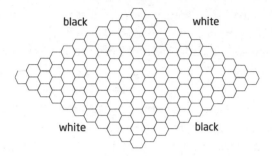

black white

white black

FIG 8 The Hex board.

mutually agreed method. A player wins by constructing a connected chain of counters joining the two edges that they own. The chain may have additional counters, side-branches, or loops, and need not be connected to other pieces of that colour. All that matters is that some series of counters forms a connected path from one edge to the opposite edge. It sounds simple, but the simplicity is deceptive. Hex is a game of deep subtlety.

Hex was first invented by Piet Hein, a Danish mathematician who is also famous for his short poetry (known as 'grooks') and numerous other offbeat ideas. He called the game Polygon, and it first saw the light of day in the Danish newspaper *Politiken* on 26 December 1942. The mathematician John Nash reinvented it independently in 1948 when he was a graduate student at Princeton. In 1969 Nash won the 'Nobel Prize' in economics, more precisely the Sveriges Riksbank Prize in Economic Sciences in Memory of Alfred Nobel. His winning idea was the concept of 'Nash equilibrium' in game theory, and his life is the subject of the brilliant biography *A Beautiful Mind*. In 2001 this was made into a movie starring Russell Crowe as Nash, which won four Oscars. At Princeton the game was known as Nash, or sometimes John – because it was often played on hexagonal bathroom tiles.[4]

In the mid-1950s Martin Gardner wrote about Hex in his Mathematical Games column, and his article is reprinted in *Mathematical Puzzles and Diversions from Scientific American*. Overnight it became a craze in virtually every mathematics department in the world. For example in 1968, when I first arrived at the University of Warwick as a graduate student, a group of us started a magazine called *Manifold*. The first issue had a Hex board drawn on the front and back covers (half on each) and an article about it between them. But it is now more than 40 years since Gardner described Hex to *Scientific American*'s readers, so I think it is time to introduce it to a new generation.

Some simple mathematical analysis illuminates the game. Since pieces are never removed, the number of moves is finite – at most 121 for the 11 × 11 board. A connected chain from edge to edge for one player necessarily blocks any connected chain from edge to edge for the other player, which makes it intuitive (but not entirely straightforward to prove) that eventually one player or the other must win. The basic point is that black, say, can only be prevented from forming a winning chain if white creates such a chain herself first.

It's an interesting challenge to prove the 'obvious' fact that if the board is filled with black and white stones, then one colour must connect two opposite edges. It's clear that both

---

[4] 'Bathroom' in the American sense, often colloquially called a 'john'.

colours cannot do this at the same time, since the chains involved must cross. It's also plausible that if, say, black stones do not connect opposite sides, then that must happen because a chain of white stones is getting in the way. However, a complete proof is less obvious. Suppose, for the sake of argument, that black's stones do not include a chain that connects the two black edges. Consider one 'component' of the black regions – all the black stones connected (by other black stones) to a black edge. Now look at the 'boundary' of this region – all the immediately adjacent white stones. Clearly, this set of white stones must connect the two white edges... but why?

Alternatively, we can prove that one player must have a winning strategy. Then the above claim easily follows. In fact, it can be proved that with proper play, the *first* player should always win. The proof, found by Nash, uses a general technique called 'strategy stealing'. Suppose, for the sake of argument, that white plays first, and there is a strategy that guarantees a win for the second player, black. If so, then white can employ unbounded brainpower to work out what that strategy is. She can then use this alleged second-player-wins strategy to *beat* black, as follows. White makes any move, and promptly forgets it. She now pretends that black is opening the game, and that she is the second player, not the first. Whichever move black makes, white plays the correct response according to the second-player strategy. There is one minor modification, however. Sometimes that strategy will require her to place a counter in the position already occupied by her first 'forgotten' move. If so, no problem: the desired cell is already occupied by a white counter, so the strategy is already complied with. She therefore makes a new move, in any unoccupied cell, and this becomes the new 'forgotten' move.

Continuing in this way, white can force a win. But now we find ourselves in a curious situation: by stealing the alleged second-player-wins strategy in this manner, white has played first – and won, no matter what moves black makes. The only way out of this logical impasse is that there never was a second-player-wins strategy. Since this game is finite and one player *must* eventually win, this implies that there must exist some first-player-wins strategy.

Observe that the second player cannot steal a first-player-wins strategy. Also, convince yourself that strategy stealing does *not* work for games like chess, where moves needed later in the strategy may be unavailable earlier on. If you can do these two things, you'll understand the proof.

At first sight this result renders the game pointless, because both players know who ought to win if they exercise perfect play. However, a similar issue arises in many other games. The most impressive example is draughts (the British name for what our American cousins call chequers), now known to be a draw if both sides play perfectly. The computer-assisted proof, organized and orchestrated by Jonathan Schaeffer, took 18 years; the main problem is the gigantic number of positions and potential lines of play. Yet rational adults remain happy to play draughts, because the perfect strategy is so complex that the human mind cannot implement it unaided. The proof that the first player should always win a game of Hex is even more elusive; it is an existence proof *only*, so the proof does not tell us an explicit winning strategy, no matter how complicated. In fact, the largest board for which a winning strategy is actually known is 9 × 9, a discovery made by Jing Yang of the University of Manitoba (see Websites). So even on a 10 × 10 board, the first player knows that in principle he ought to win, but has no idea how to go about doing it. And if that still doesn't seem fair to the second player, many people allow an optional rule: the second player may elect to swap the opening piece for one of her own, once it has been played, instead of playing on a new cell.

A full discussion of the subtlety of Hex would, and does, occupy an entire book. So I'll focus on just two features, to raise awareness of the game's subtlety. The first, which rapidly becomes clear to anyone who tries the game, is that cells do not have to be *occupied* in order to play a strategic role. Figure 9a shows a *bridge*, in which two non-adjacent cells (here occupied by black) share two cells that touch both. As long as both of the latter cells remain unoccupied by white, the two former cells are in effect already joined – for as soon as white plays on one of the intermediate cells, black

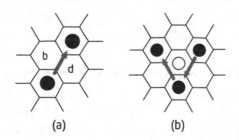

(a)   (b)

FIG 9 (a) A bridge. (b) Overlapping bridges don't work.

can play on the other. At the first level of play above rank novice, players usually attempt to build chains of bridges in the hope that their opponent won't notice. A bridge is by no means invincible, however. A black bridge can be defeated if white contrives to occupy one of its intermediate cells, whilst simultaneously threatening a winning move elsewhere. Nevertheless, this is usually far from easy, so it is best to stop your opponent building too many bridges.

A useful general principle is that a player's entire position is only as strong as its weakest link. If your opponent can attack some part of your incipient chain with good hopes of success, then you should either try to strengthen your own weakest link, or attack theirs. However, you shouldn't do this automatically on all occasions, because your opponent may notice and lay a subtle trap.

Another useful principle is to sneak up on your opponent's weaknesses from some distance away. Instead of playing a counter smack in the middle of their weak link, for instance, you can mentally map out a chain of bridges and play somewhere along that chain. By the way, don't make the mistake of forming a chain of bridges where the intermediate cells of two bridges overlap (Figure 9b), because when the opponent plays on the overlap she attacks two bridges at once. You can defend one – but not both.

Several levels of play above bridges we encounter ladders. These provide subtler opportunities and problems. A ladder arises when one player tries to form a connection to an edge but is pushed away at a fixed distance by their opponent, so that both players start forming long *forced* chains of counters, parallel to each other. Figure 10a shows the start of a ladder,

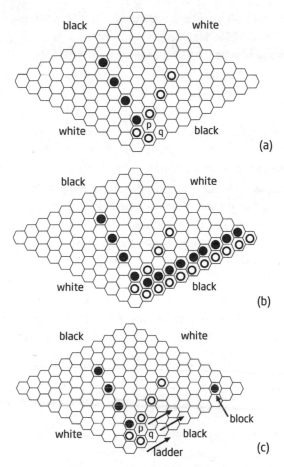

FIG 10 (a) Start of a ladder. (b) Where it leads. (c) Blocking a ladder.

with black to move. Black has no option except to play at cell *p*, otherwise white can force a win. By the same token, white now has to play at *q*. If black keeps trying to force a connection to the same edge (and for several moves she must do this or lose) then white is forced to keep blocking, and a long sequence of white counters begins to extend along the edge, with a black chain next to it. What black has failed to notice, however, is that if this process continues, white will win (Figure 10b). It is important to anticipate the occurrence of ladders, and to block your opponent's ladders before they get started. If black had an extra counter (Figure 10c) near the white edge, then black would win the ladder exchange.

Hex Strategy considers these issues, and many others, in considerable depth. It also discusses a host of variations on the basic Hex game. For example the game of Y is played on a triangular board (Figure 11) and a player wins by forming a chain that touches all three edges. The strategy-stealing proof works here, by the way, so the first player must have a

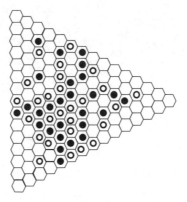

FIG 11 Board for the game of Y.

winning strategy. But again, no explicit winning strategy for player one is known, except on very small boards. Hex can be played on a map of the United States, with states acting as cells, and north–south or east–west as the edges to be connected. Here the first player can force a win by playing in California – you may enjoy working out how play then proceeds. Hex can also be played on a sphere, tiled with hexagons and pentagons. Since there are no edges, the first player to surround at least one cell (unoccupied or occupied by their opponent) wins.

WEBSITES

GENERAL:

http://en.wikipedia.org/wiki/Hex_(board_game)
http://www.swarthmore.edu/NatSci/math_stat/webspot/
  Campbell,Garikai/Hex/index.html

STRATEGIES FOR THE 7 × 7, 8 × 8, AND 9 × 9 BOARDS:

http://www.ee.umanitoba.ca/~jingyang/

PERFECT PLAY AT DRAUGHTS AND OTHER GAMES:

http://en.wikipedia.org/wiki/Perfect_play#Perfect_play

4

Jumping Champions

Prime numbers continue to puzzle the
world's mathematicians. The commonest
gap between successive primes appears
to be 6, and that's certainly true up to a
trillion or so. Given this vast quantity of
'experimental' evidence, are we justified
in concluding that 6 is *always* the
commonest gap, no matter how big the
numbers become?

MATHEMATICS IS FULL of surprises. Who would have imagined, for instance, that something as straightforward as the natural numbers 1, 2, 3, 4,…could, with minimal effort, give birth to anything as baffling as the prime numbers 2, 3, 5, 7, 11,…? The pattern of natural numbers is simple and obvious: whichever one you've got, it's easy to work out the next one. You can't say that for the primes, yet it is a simple step from natural numbers to primes: just take those that have no proper divisors.

We know a lot about the primes, including some powerful approximate formulas that provide good estimates even when exact answers aren't forthcoming. For example, the Prime Number theorem, proved in 1896 by Jacques Hadamard and (independently) Charles-Jean de la Vallée Poussin, states that the number of primes less than x is approximately $\dfrac{x}{\log x}$, where log denotes the natural (base e) logarithm. So, for instance, we know that there are roughly 4.3×10^{97} primes with fewer than 100 digits – but the exact number is a total mystery.

There are many more things about primes that we don't know. A decade ago, Andrew Odlyzko (AT&T), Michael Rubinstein (University of Texas), and Marek Wolf (Wrocław) turned their attention to the gaps between successive primes. The problem they discussed is: Up to some limit x, what is the commonest gap between successive primes? Harry L. Nelson had raised this question in the *Journal of Recreational Mathematics*. Later John Horton Conway (Princeton University) named the associated numbers *jumping champions*.

The primes up to 50 are 2, 3, 5, 7, 11, 13, 17, 19, 23, 29, 31, 37, 41, 43, 47. The sequence of gaps – the differences between each prime and the next – goes 1, 2, 2, 4, 2, 4, 2, 4, 6, 2, 6, 4, 2, 4. The number 1 appears once (and once only, since all primes except 2 are odd) and the rest are even. Here the gap 2 occurs six times, the gap 4 occurs four times, and the gap 6 occurs twice. So when $x = 50$, the commonest gap is 2, and this number is a jumping champion.

Sometimes, several gaps are equally common. For instance, when $x = 5$ the gaps are 1, 2 and each occurs once. After that, the sole jumping champion is 2 until we reach $x = 101$, when 2 and 4 are tied for the honour. After that, the jumping champion is either 2, 4, or both until $x = 179$, when 2, 4, and 6 are involved in a three-way tie. At that point the challenge from 4 and 6 dies away, and 2 reigns supreme until $x = 379$, where it is tied with 6. From $x = 389$ the jumping champion is mostly 6, occasionally tied with 2 and/or 4, but in the range $x = 491$ to 541 the jumping champion reverts to 4. From $x = 947$ onwards the sole jumping champion is 6, and a computer search shows that this continues up to at least $x = 10^{12}$.

It seems reasonable to conclude that apart from some initial competition from 1, 2, and 4, the only long-term jumping champion is 6. The computer evidence in favour seems strong. The now-defunct journal *Experimental Mathematics* was devoted to just such problems, and it was virtually unique among mathematics journals in that it existed for researchers to publish unproved conjectures obtained with the aid of computer calculations. This does not represent a weakening of the mathematical requirement of proof, because the articles clearly state that proofs are lacking. Instead, the journal's aim is to suggest interesting problems for mathematicians to answer with the usual logical rigour.

All number theorists know that there is evidence, and there is *evidence*. A pattern that persists up to numbers of a trillion or so may well change as the numbers get bigger. This problem may well be a case in point, for Odlyzko and colleagues provide a persuasive argument that somewhere near $x = 1.7427 \times 10^{35}$ the jumping champion changes from 6 to 30. They also suggest that it changes again, to 210, near $x = 10^{425}$. They support these suggestions with some non-rigorous but careful theoretical analysis, and some carefully chosen numerical experiments.

Except for 4, the conjectured jumping champions fit into an elegant pattern. This becomes obvious if we factorize them into primes:

$2 = 2$

$6 = 2 \times 3$

$30 = 2 \times 3 \times 5$

$210 = 2 \times 3 \times 5 \times 7.$

Each number is obtained by multiplying together successive primes up to some limit. These numbers are called *primorials* (like factorials but using primes), and the next few are

$$2310 = 2 \times 3 \times 5 \times 7 \times 11$$
$$30030 = 2 \times 3 \times 5 \times 7 \times 11 \times 13$$
$$510510 = 2 \times 3 \times 5 \times 7 \times 11 \times 13 \times 17$$
$$11741730 = 2 \times 3 \times 5 \times 7 \times 11 \times 13 \times 17 \times 23.$$

Oldlyzko and colleagues' main conclusion is the Jumping Champion conjecture: the jumping champions are precisely the primorials, together with 4. The basis for this suggestion is another conjecture, known as the Hardy–Littlewood k-tuple conjecture. It was stated by Godfrey Harold Hardy and John Edensor Littlewood in 1922, and it is about patterns in the gaps between primes.

Anyone who looks at the sequence of primes notices that every so often two consecutive odd numbers are prime: 5 and 7, 11 and 13, 17 and 19. The Twin Prime conjecture states that there are infinitely many such pairs. They can certainly get very large – the largest known in September 2009 are

$$65,516,468,355 \times 2^{333,333}-1 \qquad 65,516,468,355 \times 2^{333,333}+1$$

with 100,355 digits each. (As an aside: prove that twin primes always have the same number of decimal digits. If that seems obvious, here's a second challenge: if 'decimal' is replaced by 'base n', for which values of n is the analogous statement *false*?) Moreover, there is a probabilistic calculation that strongly suggests the conjecture is correct. It is based on the idea that primes occur 'at random' among the odd numbers, with a probability based on the Prime Number theorem.

Of course this is nonsense – a number is either prime or not, there isn't a probability involved – but it is plausible nonsense for this kind of problem. According to the calculation, the probability that the list of twin primes is finite is zero.

What about three consecutive odd numbers being prime? There is one example: 3, 5, 7. It is the *only* example, because given any three consecutive odd numbers, one of them is a multiple of 3 (so is not prime unless it happens to equal 3, whence the sole example). However, the patterns $p, p + 2, p + 6$ and $p, p + 4, p + 6$ cannot be ruled out by such arguments, and they seem to be quite common. For example the first pattern occurs for 11, 13, 17 and again for 41, 43, 47; later we find 881, 883, 887. You might like to work out why the pattern of final digits must always be 1, 3, 7. The second pattern occurs for 7, 11, 13, again for 37, 41, 43, and again for 877, 881, 883. This time the pattern of final digits is 7, 1, 3.

Hardy and Littlewood thought about patterns of this kind for any number of primes, and they performed the same kind of probabilistic calculation that I've just described for the twin primes. They deduced a precise formula for the number of sequences of k primes with a specified pattern of gaps, all less than some limit x. The formula is complicated to describe so I won't give it here: see the article by Odlyzko and colleagues, and references therein.

The analysis that leads to the Jumping Champion conjecture begins with the Hardy–Littlewood formula and extracts from it a formula for the number $N(x,d)$ of gaps between consecutive primes of given size $2d$, up to some limit x. We use $2d$ because the gaps have to be even, except for the gap between 2 and 3. The formula is expected to be valid

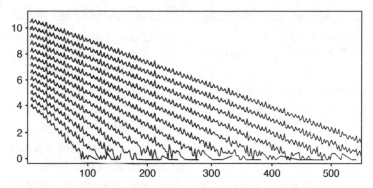

FIG 12 Plot of the logarithm of the number of occurrences of a gap of size $2d$ (vertical coordinate) against $2d$ (horizontal coordinate), for primes up to various limits x. These range from $x = 2^{20}$ (lower left) to $x = 2^{44}$ (upper right).

only when $2d$ is large and x is much larger. Figure 12 shows a plot of log $N(x,d)$ against $2d$ for $x = 2^{20}$, 2^{22},..., 2^{44}. Each graph is approximately a straight line, but with bumps. A particularly prominent bump occurs at $2d = 210$, the conjectured next jumping champion after 6. (It would look more prominent, but the logarithm flattens it out.) This kind of information suggests that the formula is not too wide of the mark.

Now, if $2d$ is going to be a jumping champion, the value of this formula has to be pretty big – at least half the number of primes less than x. The precise form of the formula (which again I won't write down) shows that the best way to achieve this is if $2d$ has a lot of distinct prime factors. It also says that $2d$ should be as small as possible subject to this condition, so the most plausible choices for $2d$ are the primorials. (The known jumping champion 4 is presumably an exception, occurring at sizes where the formula isn't a good approximation anyway.)

The conjectured formula also lets us work out roughly when a given primorial takes over from the previous one as the new jumping champion. Suppose the two primorials are $A = 2 \times 3 \times \cdots \times p$, $B = 2 \times 3 \times \cdots \times p \times q$, with p and q successive primes. Then the second takes over from the first roughly when $x = e^{A(q-1)(q-2)}$, where $e = 2.718\ldots$ is the base of natural logarithms. This is where the expected values of x for 30 and 210 to become jumping champions come from. Because of the exponential, these values of x rapidly become gigantic.

What is left to do here? Prove the Jumping Champions conjecture, of course – or disprove it, if it's wrong. If you can't do that, try something weaker: for example, prove that there exist infinitely many distinct jumping champions. In 1980 Paul Erdös and E.G. Straus proved just that, but only by assuming a quantitative version of the Hardy–Littlewood k-tuple conjecture. Unfortunately even the Twin Prime conjecture seems horrendously hard to prove, and the full Hardy–Littlewood k-tuple conjecture is almost certainly worse. More promising for recreational mathematicians is the search for other interesting properties of the gaps between primes. For example, what is the *least* common gap (that actually occurs) between consecutive primes less than a limit x? Which gap occurs closest to the average number of times – the most ordinary gap? As far as I know, these questions are wide open, even for relatively small values of x.

ANSWER

Why do twin primes always have the same number of digits in decimal notation? It may seem obvious, but the proof reveals a potential loophole, which can occur in other number bases. Let the primes concerned be p and $p + 2$. In decimal notation, it is possible for $p + 2$ to have more digits than p. However, this happens only when $p = 999...98$ or $999...99$. In the first case p is even (and at least 8) so cannot be prime. In the second case p is a multiple of 9 so cannot be prime.

The final step in the proof uses special properties of the number 10. In other bases, things work out differently. In base-n notation, p must be of the form n^k-2 or n^k-1 for some power k. That is, n^k must be either $p + 2$ or $p + 1$ for the smaller twin p of a twin prime. This can happen: for example, when $p = 3$ then n^k can be 4 or 5. The twin primes 3 and 5 (decimal) are 3 and 11 (base-4), which have different numbers of digits. In base 5, the same twin primes are 3 and 10, again with different numbers of digits.

With more effort, we can take the analysis further. If $n^k = p + 2$ then n^k is prime, so $k = 1$ and n is prime (equal to $p + 2$). If If $n^k = p + 1$ then $p = n^k-1 = (n-1)(n^{k-1} + n^{k-2} +...+ 1)$. Since p is prime, either $k = 1$ or $n = 2$. If $k = 1$ then $n = p + 1$ for the smaller twin p of a twin prime. If $n = 2$ and $k > 1$ then 2^k-1 and $2^k + 1$ must both be prime. The only case where this can occur is $2^2-1 = 3$ and $2^2 + 1 = 5$. (If 2^k-1 is prime - a so-called Mersenne prime - then it is well known and easy to prove that k must itself be prime. If $2^k + 1$ is prime - a so-called Fermat prime - then it is well known and easy to prove that k must be a power of 2. The only power of 2 that is prime is 2.)

In short: p and $p + 2$ are twin primes with different numbers of digits to base n if and only if $n = p + 1$ or $p + 2$ where p is the smaller twin prime, or $n = 2$ and $p = 3$.

FEEDBACK

'Jumping champions' was almost my last column, so there wasn't any feedback to speak of. So I'm going to cheat and tell you about a truly amazing discovery, one of the few cases where primes no longer baffle mathematicians. This is the Green–Tao theorem, proved in 2005 by Ben Green and Terence Tao. It is about patterns of primes similar to, but significantly different from, the p, $p + 2$, $p + 6$ example described a few pages back. The main result is easy to state: for any integer k there exist infinitely many arithmetic progressions of primes with k terms.

An arithmetic progression is a sequence of numbers in which each exceeds the previous one by the same fixed amount. Symbolically, such a sequence looks like

$$a, a + d, a + 2d, a + 3d, ..., a + (k-1)d$$

if it has k terms. Here d is the common difference and a is the first term. In the Green-Tao theorem, d is not specified in advance, but is constructed during the course of the proof. For many years mathematicians – often amateurs – have sought long arithmetical progressions of primes. For three terms, there is the obvious progression 3, 5, 7, in which $d = 2$. An elegant seven-term progression is

    7   157   307   457   607   757   907

with $d = 150$. But serious computer assistance is needed to find a 25-term progression, the longest one actually known (as of September 2008) is

    $6171,054,912,832,631 + 366,384 \times 23 \times d$

for $d = 0, 1, 2,..., 24$. It was discovered by Jaroslaw Wroblewski and Raanan Chermoni in 2008. Green and Tao even provided an upper limit on how big the primes need to be, in terms of k. If we write $a \hat{} b$ for a^b, this limit is

    $2\hat{}2\hat{}2\hat{}2\hat{}2\hat{}2\hat{}2\hat{}100k$.

In such expressions, the rule is to apply successive power operations $\hat{}$ from the right working to the left. So first we raise 2 to the power $100k$, then raise 2 to that power, and so on. The result is truly gigantic, and presumably a massive overestimate, but it's all we know right now, and it's astonishing that Green and Tao managed to achieve that.

By the way, any arithmetical progression of primes must be finite – they can't go on forever. But there is no specific limit that applies to them all.

It is relatively easy to extend the Green-Tao theorem to 'generalized arithmetical progressions' in which the single difference *d* is replaced by a finite list of differences, and all combinations are permitted. For instance, with two differences d_1 and d_2 we consider all numbers $a + k_1 d_1 + k_2 d_2$ with k_1 and k_2 running from 0 to some upper limit. In fact, all these numbers can be viewed as part of a longer arithmetic progression, and we just apply the Green-Tao theorem to that.

The theorem has innumerable consequences, and I mention just one: the existence of arbitrarily large magic squares composed entirely of primes (of course, these can't be consecutive integers, and they aren't even consecutive *primes*). Here's a 4 × 4 example:

```
37  83  97  41
53  61  71  73
89  67  59  43
79  47  31  101
```

The theorem says you can do this kind of thing (though using gigantic primes) for, say, magic squares of size a million, or a billion - as big as you please. For further information, see Andrew Granville's article in Further Reading.

WEBSITES

GENERAL:

http://en.wikipedia.org/w iki/Prime_number
http://mathworld.wolfram.com/PrimeNumber.html
http://primes.utm.edu/glossary/home.php

GAPS:

http://en.wikipedia.org/wiki/Prime_gap

JUMPING CHAMPIONS:

http://primes.utm.edu/glossary/page.php?sort=Jumping
Champion

GREEN-TAO THEOREM:

http://en.wikipedia.org/wiki/Green-Tao_theorem

5

Walking with Quadrupeds

Animals move in a variety of patterns, called gaits, and many of the patterns are symmetric. Now we are beginning to understand why. It all boils down to patterns in the networks of nerve cells that control animal motion. Jane and Tarzan explain.

A centipede was happy quite,
Until a frog in fun
Said, 'Pray, which leg comes after which?'
This raised her mind to such a pitch,
She lay distracted in a ditch
Considering how to run.

Mrs. Edmund Craster

TARZAN LEAPED INTO the air, kicked both legs out in front of him simultaneously, and sat down heavily on the ground. He had repeated this sequence of actions more than 20 times since Jane had started watching, and from the look on his face that was an underestimate.

It's not that Tarzan doesn't have a brain, Jane thought. *He just needs training in using it.* Indeed, she'd mapped out an ambitious education for him, and Tarzan's nose had been buried in books for weeks.

Maybe that was the problem. Jane grabbed a convenient vine and slid down.

The ape-man looked up as she approached. 'Uh, hi Jane.'

'What was all *that* about?'

'Uh – I was testing out Curie's principle.'

'Really?' It was a novel excuse.

'Yes. And it doesn't work.'

Jane gently took his hand and led him into the shade of a tree. 'Let's go somewhere cool and quiet, and you can tell me all about it.'

It took a while, but the gist was relatively simple. In one of the books that Jane had brought with her into the jungle for light reading, Tarzan had come across the statement that the human body possesses bilateral symmetry – it looks pretty much the same when reflected in a mirror. Tarzan had never seen a mirror, but he had seen the surface of a still pond, and from the pictures in the book he'd puzzled that one out. In another book, he'd come across a fundamental principle proposed by the great physicist Pierre Curie: that symmetric causes produce equally symmetric effects.

'So it seemed to me,' Tarzan said, 'that if I, a bilaterally symmetric ape – sorry, man, I keep forgetting – cause myself to walk, then Curie's principle implies that my walk should *also* be bilaterally symmetric. Which means that both legs have to move forwards together. I've been trying it ever since, but I can't seem to get anywhere. Except by sitting on my – '

'But,' said Jane, 'you've been doing it wrong. If you want a bilaterally symmetric gait, you should *hop*. Like this.' She imitated a rabbit, hopping along with both feet together, hands held like paws. Tarzan watched the spectacle in fascination. Finally he plucked up enough courage to ask what a gait was.

'It's a pattern of limb-movement, used for locomotion,' said Jane. 'Animals use all sorts of different gaits to get around. Walk, hop, gallop... Gazelles even *pronk* – they move all four legs together.'

'Hopping is all very well,' said Tarzan, 'but all it shows is that a symmetric gait is possible. My reading of Curie's principle is that *all* human gaits – in fact, all gaits of all bilaterally symmetric animals – ought to be bilaterally symmetric.' He paced thoughtfully up and down the clearing, stopping occasionally to beat his fists against his chest in frustration. 'But most of them aren't.'

Bilaterally symmetric...The same as its reflection in a mirror, thought Jane. She tried to imagine what Tarzan's walk would look like in a mirror (Figure 13). It would look like a walk. But not quite the *same* walk.

'It *almost* is,' she said. 'When you reflect a walking gait, it still looks like a walking gait.' She paused thoughtfully. 'It has to, really, otherwise people walking would look funny in a mirror. Though I suppose that's not conclusive, because letters of the alphabet *do* look funny in a mirror. Hmmm.'

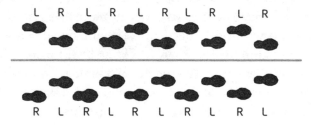

FIG 13 In the human walk, left and right feet hit the ground in turn. Reflection in a mirror (grey line) appears to swap left and right feet, which is equivalent to a time delay of half a period.

'The difference,' said Tarzan, 'is that when I put my *right* foot forward, my mirror-image puts its *left* foot – well, what appears to me to be its left foot, I don't know what its opinion is – forward. Now, on my next step, I do put my left foot forward, but by then my mirror-image is putting its right foot forward. We're always out of step with each other.'

There were times when Tarzan seemed quite bright. 'Out of *phase*, not *step*,' said Jane in excitement. 'That's why everything looks all right in a mirror. If you delay time by the amount required to take one step, then the relative positions of the legs (though not their positions on the ground) for the mirror walk looks exactly the same as they do in the original.'

'Phase?'

'Walking – like all gaits – is a *periodic* motion. It repeats at regular intervals of time. If you have two copies of the same periodic motion, but one is time-delayed relative to the other, then the fraction of the period representing the delay is called the *relative phase*. Your left leg is out of phase with your right leg by exactly half a period, that is, a relative phase of 0.5.

'Which is very interesting,' she continued, 'because it shows that gaits have symmetries in time as well as in space. After all, a symmetry is just a transformation that leaves the system looking the same afterwards as it was before. Periodicity itself is a time symmetry: shift time by one period, and everything looks the same. 'Reflect left/right *and* shift phase by 0.5' is a mixed spatio-temporal symmetry of the human walk. Isn't that grand?'

'What was the relative phase when you were hopping? Was it 0?' asked Tarzan tentatively.

FIG 14 Eight snapshots of the bound of a kangaroo. The animal's bilateral symmetry is maintained at all times.

'Exactly. The two legs moved together, so there was no difference in phase. The same goes for a kangaroo when it hops' (Figure 14).

'What's a kangaroo?'

'Oh, sorry – there aren't any in Africa, they live in Australia. They hop around on two legs.'

The ape-man leaped to his feet, performed a curious war-dance, and crashed to the ground. 'I was trying to get a relative phase of 0.3,' he explained.

'I'm not sure you can,' said Jane.

'Of course I can! All I have to do is make my left foot lag behind my right by 0.3 of a period!'

'True.'

'But that seems hard.'

'Maybe it's because it isn't a true symmetry,' said Jane. 'You see, if *everything* looks the same after swapping left and right and shifting phase by 0.3, then not only must your left leg be 0.3 out of phase with your right, but your right must also be 0.3 out of phase with your left. So the right leg is 0.3 + 0.3 = 0.6 out of phase with itself, which is silly.'

'Dangerous, too,' said Tarzan, ruefully rubbing his legs.

'Hey! There's a theorem in all this!' Jane yelled. Fans of Edgar Rice Burroughs will recall that Jane's father was Professor Archimedes Q. Porter, so it is unsurprising that his daughter should have inherited some of the family's mathematical ability. 'If left–right reflection combines with a phase shift to give a symmetry,' Jane went on, 'then the phase shift must be either 0 or 0.5. Nothing else is possible.'

'Why?'

'Because the same argument applies. If each leg is delayed relative to the other by some phase, then each leg is delayed relative to itself by twice that phase. Now, it's possible for a leg to be delayed relative to itself – but only by an integer multiple of the period, because that's effectively the same as no delay at all. So twice the phase shift is 0, 1, 2, 3, and so on; which implies that the phase shift is 0, 0.5, 1, 1.5, and so on. But 1 has the same effect as 0, and 1.5 the same effect as 0.5, because of periodicity.

'Which means,' she continued, 'that the gaits of two-legged animals can have only those two symmetries. Apart from no symmetry at all. I wonder if that can actually – ' Tarzan limped towards her, dragging one leg. 'That's it, exactly! You do catch on quickly, Tarzan.'

He squatted next to her, rummaging through the hair on his chest in search of little bits of salt until Jane slapped his wrist. 'Four-legged animals must be more complicated,' he said.

'True. There are lots of quadruped gaits.' Figure 15 shows the eight most common. The *bound* has left–right symmetry, like the two-legged hop. The *pace*, common in giraffes (Figure 16) and camels, is like the human walk: it changes phase by half the period if left and right are swapped.

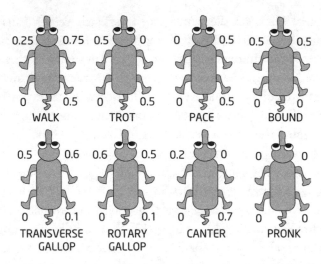

FIG 15 Eight common quadruped gaits, showing the relative phases of the legs.

FIG 16 The walk of a giraffe, which breaks bilateral symmetry. The second four frames are the same as the first four, but reflected left–right (with respect to the *giraffe*, not the page).

'What I don't understand,' Tarzan mused, 'is why Curie's principle doesn't work. Why are gaits less symmetric than the whole animal?' At that moment, Heftilump the elephant ambled through the glade, trumpeting his pleasure at seeing Tarzan. Tarzan trumpeted back. 'Mind you,' he continued, 'I don't think a pronking elephant bears contemplation. Survival of the flattest...it would never have evolved.'

'Symmetry-breaking,' said Jane. 'That's why Curie's principle fails.'

'What's symmetry-breaking?'

'It happens when a symmetric system behaves in a less symmetric way.'

'Oh. You mean, it's what happens when Curie's principle fails.'

'Precisely!'

'So...Curie's principle fails whenever Curie's principle fails. Great. That really clarifies the issue, Jane.'

Jane growled like an angry lioness. *Damn! Now he's got me doing it!* 'The important point to understand, Tarzan, is that Curie's principle *can* fail. Let me show you how. Where's Jim?'

Young Jim Pansy was always hanging around near the hut – usually in it, stealing bananas – and Jane collared the beast with ease. She tied a knot in the end of a vine, and sat the young ape on it, where it clung, chittering excitedly until she stuffed a banana in its mouth to shut it up.

'When Jim sits still and the vine hangs vertically downwards,' said Jane didactically, 'the entire system has circular symmetry.' Tarzan looked baffled. 'I mean, if you walk round it, it looks pretty much the same from all directions.' Tarzan inspected Jim's face, then walked round to the far side. He looked more baffled than ever. 'You have to pretend Jim is a featureless spherical lump, Tarzan.' He nodded happily.

'Now, suppose I grab the vine where it's draped over this branch, and pull it up and down *gently* like this...Then Jim bobs up and down, but he doesn't move sideways. The important part of the system, the bit of vine hanging from the branch with Jim attached, is still circularly symmetric, even though it bobs up and down; but look what happens.' As Jane pumped the vine more energetically, Jim began to swing in an arc, short at first, then longer and longer. The chimp squealed in delight, waved its arms, and fell off, terminating the experiment.

'I saw it,' said Tarzan, 'but I'm not sure what I saw.'

'Symmetry-breaking,' said Jane. 'The perfectly symmetric state of the system is to hang vertically. But, when I pump it, that state becomes *unstable*. It still exists, mathematically, but you don't observe it in practice because any tiny random deviation tends to grow. Since the symmetric state can't occur, then naturally the system has to do something else, which perforce has to be less symmetric.'

'Ah.' He paused. 'What does "perforce" mean?'

Jane ignored him. 'However, it's not *totally* asymmetric. Jim was swinging to and fro in a plane. If you think of that plane as a mirror, then his swing is symmetric under reflection in that mirror. That's an example of a *standing wave*.

'But that's not all.' She picked Jim up, stuffed another banana into his mouth to mollify him, and attached him to the vine again. 'There's another type of periodic oscillation that Jim can perform, too.' She gave the ape a shove, and he swung in circles. 'Now, you might think that this motion has circular symmetry, but that's not true. If you rotate the system through some angle, then it doesn't look *exactly* the same.'

'No, it's like the walk in a mirror. It's the same general kind of motion, but in a different place at a given time.'

'Right. What does that mean?'

'Much the same, but the timing's wrong…of course. It's a phase shift again.'

'You've got it. If you rotate the system, and apply a suitable time delay, it looks *exactly* the same as before. And in this case the time delay is the same as the rotation, in the sense that a rotation of 0.4 of a turn needs a time delay of 0.4 of a period, and so on. That's called a *rotating wave*.'

'Let me run this up an acacia tree and see who gets scratched,' said Tarzan. Jane began to wish she hadn't included a book about business in her travelling library. 'When the perfectly symmetric state becomes unstable, the symmetry can break either to a standing wave, or a rotating wave. The standing wave has a purely spatial symmetry – reflection in its plane. The rotating wave has a mixed spatio-temporal symmetry.'

'That's it, exactly!' Tarzan beat his chest and howled in triumph, while Jane shook her head. It wouldn't go down well in the House of Commons; the ape-man's education still had some way to go. 'However, the circular symmetry hasn't *totally* vanished.' She grabbed the vine. Jim looked worried. 'Choose a vertical plane.'

'In line with that monkey-puzzle tree,' said Tarzan. Jane gave Jim a push in that direction; the ape oscillated to and fro in the plane that Tarzan had chosen. 'Which planes will that work for?'

'Any of them, I guess,' said the ape-man. 'Provided they're vertical and run through the point where the vine runs over the branch.'

'Right. Planes through the symmetry-axis. And how are those planes related?'

'Hmmm...They're all rotations of each other. I see! Instead of having a single state of the system, unchanged by all rotations – that is, a fully symmetric state – you get *lots* of less symmetric states, all related *to each other* by rotations.'

'Exactly. The whole set of motions still has circular symmetry, in the sense that if you rotate any motion, you get another one in that set. But it may not be the one you started with. The symmetry isn't so much *broken* as *shared*.'

At that moment a spotted orange shape shot across the clearing, yowling, collided with Tarzan, and they fell in a struggling heap. There was a brief scuffle, from which the ape-man emerged wearing a broad smile, and cradling a large cheetah. 'Look, Spot's come to visit!'

'Yes, and using what I judge to have been a transverse gallop,' said Jane, 'which is one of the least symmetric gaits' (Figure 17).

FIG 17 The transverse gallop of a cheetah.

'What symmetry does it have?' asked Tarzan.

'You can read it off from the phase shifts,' said Jane (see Figure 15). 'In the transverse gallop, diagonally opposite legs are 0.5 out of phase. There's also a curious phase lag of about 0.1 between the front left and front right legs, which I'm not going to explain because then it will get *really* technical. It's probably related to the efficient use of energy by the animal. Anyway, the symmetry is this: *interchange diagonal pairs of legs and shift phase by half a period.*'

'I wonder what kind of symmetry-breaking could create that kind of motion?' said Tarzan. But the Sun was setting. They retired to their hut.

Next morning, Jane was awoken by a tremendous screeching and chattering, like a pack of monkeys. When she looked down into the clearing, that's pretty much what she saw. Tarzan had rigged up a complicated network of vines between four trees (Figure 18) and was trying to use bananas

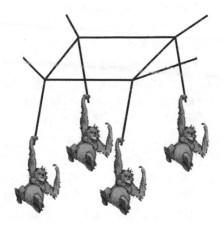

FIG 18 Tarzan's Central Pattern Generator simulation.

to bribe some young chimps to cling to the ends of four hanging vines. *Apes, not monkeys. Same difference.*

'It's a model of what the biologists call a Central Pattern Generator,' said Tarzan happily. 'I've been doing some more reading. Each chimp represents a component of the animal's neural circuitry, controlling a leg. The vines are interconnections that couple the neurons together, so that they affect each other. The dynamics of the circuit controls the rhythms of the gait. Look!' He gave one chimp a shove and it began to swing; the impulses transmitted along the linking vines soon set the other chimps swinging in sympathy. A rather complex pattern was just setting in when one chimp jumped off to steal another's banana.

'Just a hardware problem,' said the ape-man, picking up the miscreant and replacing him on his vine. 'The basic concept is OK. Each network permits a whole range of oscillations. That's why a single animal can employ several different gaits, depending on speed, terrain, and so on. I can get most of the standard gaits using a square arrangement. Oddly enough, the one that I can't seem to get is the walk. That's a kind of figure-8 rotating wave, in which the front left, back right, front right, back left legs move in sequence, with 0.25 phase lags. But I *can* get that if I rearrange the vines to make two of the side-connections cross.'

'Let me see if I understand what you're suggesting,' said Jane. 'You're looking at various networks of coupled oscillators, and finding out what kinds of symmetry-breaking can occur. Then you're trying to match the results up with actual gaits, on the assumption that each leg is controlled by one oscillator.'

'Well, of course. I mean, *anyone* could see that. Though each "oscillator" could be a complicated circuit in practice. The point is, it works! Look, suppose you want a bound. Then you set the front two "legs" moving together, and at the same time' – he rushed to the other end of the clearing – 'you set the other two legs going together, but 0.5 out of phase. Of course, you can *start* the chimps swinging in any pattern you like; but only a few patterns persist for very long. The rest get all muddled up. So I figure those are the natural oscillation patterns of the network. It's just as easy to get the trot, the pace, and the pronk.

'The two types of gallop aren't so much harder, but I'm having real trouble persuading these chimps to canter, I can tell you! Probably just need more bananas to iron out the bugs.'

'Tarzan, aside from the appalling mixed metaphors, that's really rather impressive – ' Jane began, but the ape-man had dived into the bushes, shouting. 'Bugs! Bugs! It ought to work for bugs too!' He reappeared waving a large green beetle, and placed it on a rock. After a hesitant start, the insect scuttled off.

'Tripod gait,' said Jane. 'Legs go together in threes, one triple being 0.5 out of phase with the other (Figure 19). Front and back one side, middle on the other. Nice symmetries.'

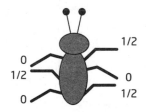

FIG 19 Tripod gait of an insect.

By late afternoon Tarzan had rigged up a set of six vines linked in a hexagon, and six puzzled chimps were swinging happily in a tripod gait. Alternate chimps swung inwards and outwards, 0.5 out of phase.

As she dropped off to sleep that night, Jane found herself thinking *I hope Tarzan doesn't start wondering about*…but she fell asleep before she could complete the thought.

Just after sunrise, she was awoken by the sound of huge trees crashing to the ground, against a background of the most appalling screeching she'd ever heard. Tarzan was extending the clearing to make a long track. A huge pile of vines lay along both sides, a heap of bananas as large as their hut at one end, and chimpanzees were charging around everywhere. She tried to count them. There must have been at least a hundred.

Exactly a hundred, of course. Her thought of the previous evening completed itself. *I hope Tarzan doesn't start wondering about centipedes.* Not that centipedes actually have a hundred legs, but then, Tarzan was very literal-minded.

A new thought occurred to Jane. *Oh my God. I just hope nothing reminds him about millipedes.*

WEBSITES

GENERAL:

http://en.wikipedia.org/wiki/Animal_locomotion
http://en.wikipedia.org/wiki/Terrestrial_locomotion_in_
  animals

HORSES:

http://en.wikipedia.org/wiki/Horse_gait

INSECTS:

http://www.mindcreators.com/InsectLocomotion.htm

EARLY PHOTOGRAPHS, INCLUDING MOVIES:

http://commons.wikimedia.org/wiki/Category:Eadweard_
  Muybridge
http://en.wikipedia.org/wiki/Eadweard_Muybridge

6

Tiling Space with Knots

Square tiles, rectangular tiles, hexagonal tiles, curved tiles - mathematicians have been charmed by their patterns, startled by their versatility, and baffled by apparently simple questions that turn out to be amazingly hard. But have you ever thought about *knotted* tiles?

S HAPES THAT TILE the plane – filling it completely without overlapping – are a recurring theme in both recreational and mainstream mathematics. Solids that 'tile' three-dimensional space have also attracted a lot of attention. In fact, so many people have worked on these questions that it would be easy to imagine that nothing new remains to be done. That this is definitely not so was brought home to me by a beautiful article in the *Mathematical Intelligencer* by Colin C. Adams (Williams College). Adams has discovered general methods for creating three-dimensional tiles with highly intricate topology; in particular they can be knots.

All of Adams's three-dimensional tilings are constructed from congruent copies of one single shape, called the *prototile*. The simplest three-dimensional tiling uses a cube as a prototile, stacking the cubes like a three-dimensional chequerboard. This 'cubic lattice' tiling might seem prosaic, but simple modifications can create tiles with a surprisingly complex topology, as we'll see.

Topology is 'rubber sheet geometry', the geometry of continuous transformations; that is, it studies those properties of shapes that remain unchanged when the shape is stretched, squashed, bent, twisted, or generally deformed in a continuous manner (no tearing or cutting). Such a deformation is called a topological equivalence: for example a cube is topologically equivalent to a sphere – just round off the corners. Topological properties include fundamental concepts such as connectedness and knottedness.

A favourite shape for topologists is the torus, shaped like a doughnut or an automobile tyre. For the purposes of this article I'm thinking of a solid torus – the dough of the doughnut and not just the sugary surface. To get your mind moving along topological lines, you should begin by inventing a prototile that is topologically equivalent to a torus. Think about it before reading on. Figure 20a shows one possible solution. The prototile is a cube with a square hole bored through the middle. Two 'lugs', with the same cross-section as the hole, are placed at the middle of opposite faces; each lug is half the length of the hole.

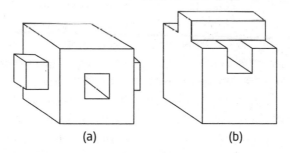

(a)   (b)

FIG 20 (a) Toroidal tile formed by boring a hole through a cube and adding matching lugs. (b) An alternative way to make a toroidal tile.

Topologically, this prototile is just a solid torus: if you made it from modelling clay you could squash the lugs flat and then round off the corners to get the traditional doughnut. You can build a flat slab one cube thick from copies of this prototile by placing them like the squares of a chessboard, with those corresponding to a black square oriented at right angles to those corresponding to a white square, so that the lugs fit neatly into – and fill – the holes. Then you just stack a pile of slabs on top of each other.

With this prototile you could make real tiles from wood and actually fit them together one by one: they tile space but do not interlock. An alternative, shown in Figure 20b, involves prototiles that interlock. From now on we'll allow prototiles to interlock: we're looking for mathematical patterns that tile space, but we're not worried about how to assemble them from separate tiles.

Both solutions illustrate the 'pick-and-mix' principle, which can be seen most clearly in the plane (Figure 21). Start with a simple tiling – here squares. Subdivide each tile into several pieces, using the same subdivision in each tile. Now assemble a new prototile by choosing one copy of each piece – but not necessarily from the original square. The result automatically tiles the plane. Similar constructions apply to three-dimensional space. As a variation, the original simple tiling may involve placing the prototile in different orientations in some regular manner. The tilings described in the figure can be viewed as applications of the pick-and-mix principle to the tiling of space by a cubic lattice. For Figure 20a the basic cube is divided into three pieces: a cube with a tunnel, and two lugs that split the tunnel into equal

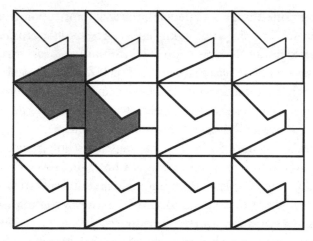

FIG 21 Pick-and-mix principle, here illustrated for a tiling of the plane by squares. Subdivide each square into several pieces; then create a prototile that uses one copy of each piece but taken from more than one component square.

halves. For Figure 20b it is divided into a cube with a square groove, and a rectangular box that fills the groove. If these cubes are stacked in a lattice with suitable orientations, and the appropriate pieces are then reassigned to neighbouring cubes, you get the tilings just described.

The 'lug-and-hole' construction can easily be modified so that the torus has more than one hole: bore several parallel holes in a neat row, with matching pairs of lugs, each half as long. Indeed the same basic idea leads to tiles with more exotic topology, known as 'cubes with holes'. To obtain a cube-with-holes start with a solid cube, and bore several tunnels through it, always starting at the top face and ending at the bottom face. These tunnels can wind round each other, form knotted loops, and generally intertwine in topologically intricate ways. Any cube-with-holes can be modified to create a topologically equivalent prototile. Simply split each tunnel in two, and add lugs to the left and right faces of the cube to correspond to the appropriate half-tunnels. These prototiles fit together in exactly the same way as those in Figure 20a: again we are applying the pick-and-mix principle.

Moreover, the addition of the lugs does not change the topology of the original cube-with-holes, because you can imagine each lug growing continuously outwards from the face to which it is attached. Call this the 'sprouting principle' – a shape retains the same topology if it sprouts extra protuberances. There is one important restriction: the protuberances must not themselves develop holes, because this is not a continuous transformation. To be precise, the protuberances must be topologically equivalent to a cube, and they must be attached at only one face of that cube. (To a topologist, a long thin wiggly tube attached at one end is equivalent to a cube attached at one face.)

This is quite a general idea, but many interesting topological shapes are not equivalent to a cube-with-holes. In order to deal with them, Adams introduces another, much cleverer technique. I'll illustrate this using a solid torus tied in a simple overhand (or trefoil) knot; a very similar method works with any knot whatsoever. The basic idea is to think about how you might cast a trefoil knot in bronze using a mould whose pieces fit together to make a cube. Then you apply the pick-and-mix principle. In order to retain the topology of the knot, it turns out that the pieces of the mould must be topological cubes.

Figure 22a shows just such a mould. Two of the pieces are half-cubes with indentations along one face, and the third is a strange tree-like structure. The role of the tree is to join overlapping regions of the knot together to convert it into a many-holed torus. It is formed from three squarish patches, which glue up the overlaps, and these are piped together by thin tubes – so that only one extra piece is needed instead of three, and that piece is topologically equivalent to a cube. The top and bottom pieces of the mould fit together to form a normal square-sided cube, except for a region that corresponds precisely to the knot plus the tree. The stem of the tree extends to the edge of the overall cube.

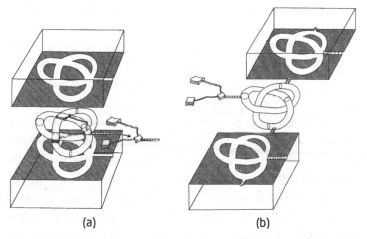

(a)   (b)

FIG 22 (a) Casting a trefoil knot using a three-part mould that fits together to make a cube. (b) A prototile formed from this by applying the pick-and-mix principle. Despite its complicated appearance, the sprouting principle implies that it is topologically equivalent to the trefoil knot.

Why introduce the extra complexity of the tree? The reason is that you cannot cast a trefoil knot from a mould with only two pieces, if those pieces are to be topologically equivalent to cubes. The tree converts the knot into a shape that can be cast in this manner.

Having constructed the three-part mould, you now use the pick-and-mix principle to create the prototile shown in Figure 22b. Begin with a cubic lattice whose cubes are split into four pieces: a trefoil knot plus its three-piece mould, as just described. Imagine space filled with such cubes, arranged in a cubic lattice. Then choose one copy of each piece as shown in Figure 22b: the knot from one cube, the top half-cube from the one behind it, the bottom half-cube from the one in front of it, and the tree from the one to its left. You must also cut a few grooves and add matching tubes with semicircular cross-sections, as shown, so that the pieces fit together into a single – rather elaborate – prototile. Despite its curious spindly architecture, this prototile is topologically equivalent to the original trefoil knot. This follows from the sprouting principle, because the prototile is formed by adding three protuberances to the trefoil knot, and despite their complex shapes, those protuberances are topologically equivalent to cubes.

This method, though topologically elegant, leads to rather complex shapes, and you could be forgiven for wanting shapes more like an ordinary knotted tube. Adams has an answer to that too: he starts with a cube and cuts it into congruent knotted pieces. Figure 23 illustrates such a decomposition into four symmetrically related trefoil knots. If you start with a cubic lattice, and break each cube into four trefoil

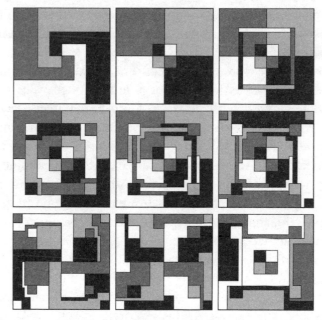

FIG 23 Successive slices through a cube composed of four symmetrically placed trefoil knots. To form the knots, stack the slices on top of
each other and glue adjacent regions with the same colour together.

knots in the manner shown, then you have tiled space with
trefoil knots.

There are many unsolved problems about knotted tiles:
here's one that's suitable for recreational mathematicians.
Suppose you start with a cube and subdivide it into n^3 smaller
cubes, each $1/n$ the size, in the obvious way. Now colour
those sub-cubes with four colours so that the sub-cubes of a
given colour form a shape topologically equivalent to a trefoil
knot. Do this so that the four knots are symmetrically related

by right-angle rotations, just as they are in Figure 23. The question is: what is the smallest value of n for which this can be done?

By covering each square in Figure 23 by a suitably fine grid you can convince yourself that a large enough value of n works – I leave you the pleasure of finding the precise value. The unsolved problem is whether there exist similar diagrams based upon a smaller grid. Note that n must be even if the four knots are to be symmetrically related, and it is fairly easy to rule out small values of n. You might also care to investigate whether the minimal value for n can be reduced by omitting the symmetry condition.

FEEDBACK

Michael Harman, a chartered patent agent living in Camberley in the UK, sent me a long letter describing several novel approaches to finding knotted tiles. An especially interesting idea is to start with a 'torus knot', formed by winding a length of string around a solid torus (Figure 24). Several congruent copies of such a string tile the surface of the torus, and this tiling can be extended to fill the interior, with the tiles remaining congruent.

It is well known that a cube can be dissected into two congruent tori (plural of 'torus'). Harman observes that each

FIG 24 Torus knot. This one wraps eight times through the hole for every three turns round the entire torus.

of these can be dissected into two congruent knots, so we obtain a new dissection of a cube into four congruent knots. He adds 'it is also worth noting that the dissections of the two tori can be either directly matching or mirror-images of each other.'

WEBSITES

GENERAL:

http://www.scienceu.com/geometry/articles/tiling/
http://mathworld.wolfram.com/Tiling.html
http://en.wikipedia.org/wiki/Tessellation

NON-PERIODIC TILINGS:

http://en.wikipedia.org/wiki/Penrose_tiling

THREE-DIMENSIONAL TILINGS:

  http://en.wikipedia.org/wiki/Convex_uniform_honeycomb

TORUS KNOTS:

  http://en.wikipedia.org/wiki/Torus_knot
  http://mathworld.wolfram.com/TorusKnot.html

TABLES OF KNOTS:

  http://www.math.toronto.edu/~drorbn/KAtlas/Knots/

KNOT INVARIANTS:

  http://en.wikipedia.org/wiki/Knot_theory

7

Forward to the Future 1: Trapped in Time!

Time travel has been a theme in science fiction ever since H.G. Wells wrote *The Time Machine*, just over a century ago. For the last few decades it has also been a theme in relativistic physics. Despite the many paradoxes, the laws of physics as currently understood do not seem to rule out travelling through time. Welcome to Hawkrose & Penking Heavy Engineering.

I **WAS JUST** finishing my shift at Hawkrose & Penking Heavy Engineering when I heard a faint whining noise. It seemed to be coming from the virtual reality simulation area. The place gets pretty dead late in the evening, and I was the only person around. I had little choice but to find out what it was, but I was nervous, I can tell you. It could have been a cyberspatial break-in. Physical security is unbeatable in the year 3001 – we have DNA-sensitive robot guards, for example – but electronic security is another thing altogether. There are just too many smart crooks with electronics training.

The room was full of acrid smoke. It had to be a physical break-in, which was impossible. I started to sweat. The smoke began to clear.

There was a strange contraption in the middle of the room, a delicate framework of shiny metal, glass, and what appeared to be off-white plastic. It had an old-fashioned look. A man was sitting in the middle of it, hidden inside a black cloak. He moved.

'*Security!*' I shouted. 'This room is sealed. Come out with your hands raised. Do not touch any lasers, phasers, rocket-launchers, or other weaponry, or you will be instantly annihilated by our biocybernetic defence systems.' I was bluffing, but maybe he wouldn't know that. He climbed out. 'Identification?' I asked.

'Uh – you wish to be apprised of my name, sir?'

He sounded polite, and *very* old-fashioned. What was he trying to pull? 'Identify yourself immediately,' I said.

'You may call me the Time Traveller. I am a friend of Mr. Herbert Wells.'

Herbert – wait, did he mean Herbert *George* Wells? H.G. Wells, the famous science fiction writer? 'Yeah, and singularities grow on trees.' I spread him out against a wall and searched him. I found some very strange items, including a quill pen. I looked closely at the machine. It was made of steel, tin, glass, and crystal, with beautifully engineered brass fittings. Some parts were made of a white plastic material, which I couldn't place.

I knew his story made no sense – but somehow it sounded convincing. There was a kind of *ancient* feel to the equipment, a genuine antique. You can't fool me when it comes to engineering.

'Suppose that on a whim I pretend to believe you,' I said. 'How did you get here and why?'

'I had no choice. I was on my way to the distant future when I smelt smoke. I turned off the machine, but too late. The temporal selection gear has stripped its teeth.' He fiddled inside the machine for a moment and pulled out a very sad-looking disc of the plastic stuff, a wisp of smoke still rising from it. 'Perhaps you could be so kind as to make me a new one?'

'That depends,' I said. 'On what sort of plastic you need.'

'Excuse me, good sir, but what is 'plastic'?'

He was either a very good actor or he was telling the truth. He didn't know what plastic was. I said 'White stuff, like that.'

'Oh, this? This is elephant ivory. It is the only material fit for the purpose, something to do with its animal origins. But it must still be very common.'

At that moment I became convinced. *Nobody* in 3001 could get hold of ivory. For one thing, trade in the stuff had been prohibited for a thousand years. For another, the last elephant had been slaughtered by poachers nine hundred and fifty years ago. What ivory remained was in museums, priceless, and had aged to a dull yellow.

This stuff was *fresh*.

'Not a hope,' I said, explaining why I couldn't get the materials needed to make a new gearwheel.

The Time Traveller looked close to tears. 'Then I am trapped,' he whispered.

'Maybe, maybe not,' I said. 'If there's a way, Hawkrose & Penking will find it. Now, tell me how this contraption works, and I'll see what we can come up with.'

He made a visible effort and pulled himself together. 'You may recall that in the 1894–5 issue of *The New Review* my friend Mr. Wells published a story called "The Time Machine".'

As it happened, I did: my hobby is ancient literary history. I've always felt it was fitting that the magazine couldn't decide which year to be published in.

'The tale was inspired by a true invention,' the Time Traveller continued, when I nodded. Mr. Wells himself explained the main idea when he wrote that "There is no difference between Time and any other of the three dimensions of Space except that our consciousness moves along it." This machine moves in a different direction from our consciousness, that's all. When it works.'

'Interesting,' I said. 'Not entirely true, but interesting.'

'Not entirely true?' So I had to explain to him some basic relativity, the kind kids get in the gestation-tanks before being decanted. Starting with Special Relativity.

'The main thing to remember,' I said, 'is that "relativity" is a silly name.'

'Then why do you employ it?'

'Historical accident. We're stuck with it. Unless you can get your machine working, go back, and persuade old Albert to invent a better one.'

I explained that the whole point of Special Relativity is *not* that 'everything is relative', but that one particular thing – the speed of light – is unexpectedly *absolute*. If you're travelling in a car at 50 kph (kilometres per hour) and you fire a gun forwards, so that the bullet moves at 500 kph relative to the car, then it will hit a stationary target at a speed of 550 kph, adding the two components (Figure 25a). However, if instead of firing the gun you switch on a torch, which 'fires' light at a speed of 1,079,252,848 kph, then that light will not hit the stationary target at a speed of 1,079,252,898 kph (note the final 98, not 48). It will hit it at 1,079,252,848 kph, exactly the same speed that it would have had if the car had been stationary (Figure 25b).

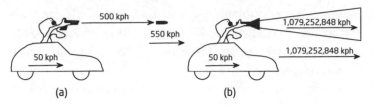

FIG 25 (a) In Newtonian mechanics, relative velocities combine by addition. (b) In relativistic mechanics, the speed of light is constant.

'You can prove this for yourself,' I told him. 'You need a shoebox, a torch, and a mirror.'

'Torch?'

'Oh, heck – use a lantern. Cut a small hole in the front of the box, to let the light in. Cut a flap in the top so that you can open the box and look inside; and write 'THE SPEED OF LIGHT IS 1,079,252,849 KPH' on the bottom of the inside of the box. Stand still, close the flap, aim the lantern at the mirror so that the beam reflects back into the box through the hole, and open the flap to read off the speed of light. Then *run towards the mirror* and repeat the experiment. Funny, you get 1,079,252,849 kph both times…'

'That,' said the Time Traveller haughtily, 'is an extraordinarily silly experiment.'

'True. But with more sophisticated equipment *you get the same answer* – as Albert Michelson and Edward Morley discovered between 1881 and 1894. They were trying to detect the motion of the Earth relative to the "ether", the all-pervading fluid that was thought to transmit all electromagnetic radiation, light included. If Newtonian physics were correct, that motion would show up as a difference in the apparent speed of light when the Earth was at opposite points of its orbit, moving in opposite directions. But they couldn't find any difference in the speed at all, even with very sensitive equipment.'

'Yes, I know about their work. It seemed to me that all it proved was that the Earth must carry the ether along with it when it moves in orbit.'

That had never occurred to me, though doubtless others had thought of it at the time, and presumably it had been dismissed for good reason. I improvised. 'It's a cute theory.'

'Cute?'

'Uh – clever. But you'd expect to see funny effects in the light from distant stars if the ether was swirling around like that. Michelson and Morley came to the conclusion that either there isn't an ether at all, or the Earth *isn't* moving relative to it – which is not very credible – or that there's something pretty weird about light.'

'And which of those alternatives it true?'

'Well, a physicist called Albert Einstein is generally credited with the theory – called Special Relativity, like I said – that there's something pretty weird about light. He published it in 1905. But a lot of other people – among them Hendrik Lorentz and Henri Poincaré – were working on the same idea, because it was widely recognized that Maxwell's equations for electromagnetism didn't entirely fit with Newtonian mechanics. The problem was one of "moving frames of reference". How do the equations change when the observer is moving? There are formulas that answer this question. In Newtonian mechanics, for example, velocities measured by (or relative to) a moving observer change by subtracting the motion of the observer. But Newtonian transformations can mess up Maxwell's equations. The answer is to use different formulas, called Lorentz transformations. They keep the speed of light constant, but have spin-off effects on space, time, and mass. Objects shrink as they approach the speed of light, time slows down to a crawl, and mass becomes infinite.'

'It is difficult to credit such a strange tale.'

'You arrive in the middle of this building in what you claim is a time machine, and you say *I'm* telling an incredible story?'

'Well, when I started out, this building did not exist. In any case, sir, I am here.'

'Yes. And so is Special Relativity. Now, I admit it's not easy to think about this kind of thing using just the formulas, and the idea didn't really take off until 1908 when the mathematician Hermann Minkowski provided a good geometric model for relativity – a simple way to *visualize* it – now called Minkowski (or *flat*) space-time.

'Precisely *because* relativity is about the non-relative behaviour of light, everything in it depends heavily upon which "frame of reference" is used by an observer. Moving and static observers see the same events in different ways.'

'That I understand. The time machine works on just such a principle.'

'Yeah, right. But you're thinking in Newtonian physics. OK, there's a lot in common. Mathematically, a frame of reference is a coordinate system. Newtonian physics provides space with three fixed coordinates (x,y,z). The structure of space was thought to be independent of time, and it was not

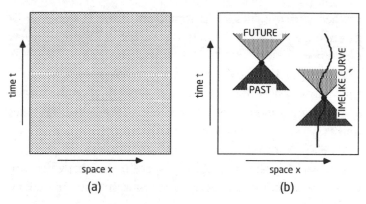

FIG 26 Minkowski space-time. (a) Space-time coordinates. (b) Light cones and a timelike curve.

traditional to represent time as a coordinate at all. Minkowski introduced time as an explicit extra coordinate. We can draw two-dimensional Minkowski space-time as a plane (Figure 26a). The horizontal coordinate, x, determines a particle's position in space; the vertical coordinate, t, determines its position in time.'

'But that is what I told you!' the Time Traveller said excitedly. 'Time is just a fourth dimension!'

'Yes, but there's an extra wrinkle that your civilization didn't know about. I'll get to it in a moment, but first I have to explain something about my drawings. In full-blooded Minkowski space-time x is three-dimensional; but for convenience let's pretend it's one-dimensional. Later on I'll have to represent space as being two-dimensional. The problem is that four dimensions of space-time don't fit conveniently onto two-dimensional paper, so a lot of the mathematics involves tricks for cutting down the number of dimensions of space. The simplest trick is to ignore a few dimensions.

'As the particle moves, it traces out a curve in space-time called its *world-line*. If the velocity is constant, then the world-line is straight, and its slope depends on the speed. Particles that move very slowly cover a small amount of space in a lot of time, so their world-lines are close to the vertical; particles that move very fast cover a lot of space in very little time, so their world-lines are nearly horizontal. In between, at an angle of 45°, are the world-lines of particles that cover a given amount of space in the same amount of time – measured in the right units. Those units are chosen to correspond via the speed of light – say years for time and light-years for space. What covers one light-year of space in one year of time?'

'Um – light?'

'Of course. So 45° world-lines correspond to particles of light – light rays or photons – or anything else that can move at the same speed.'

'Particles of light?'

'Look, just accept it as an image, OK? Think of light rays, if it makes you feel more comfortable.'

'As you wish. My head is starting to ache.'

'You ain't seen nuthin' yet, buster.'

'My name is not Buster.'

'A figure of speech. Anyway, you haven't told me your name. Now, the extra wrinkle is that relativity forbids bodies that move faster than light. The mathematical reason is that their lengths would become imaginary – involving the square root of minus one – as would masses and the local passage of time. So the world-line of a real particle can never slope more than 45° away from the vertical. Such a world-line is called a *time-like curve* (Figure 26b). Any event – point in space-time – has associated with it a *light-cone*, formed by the two diagonal lines at 45° inclinations that pass through it. It's called a cone because when space has two dimensions, the corresponding surface really is a (double) cone. The forward region contains the *future* of the event, all the points in space-time that it could possibly influence; the backward region is its *past*, the events that could possibly influence *it*. Everything else is forbidden territory, elsewheres and elsewhens that have no possible causal connections with the chosen event.

'Now, Pythagoras' theorem tells us that in ordinary space, the distance between two points with coordinates (x,y,z) and (X,Y,Z) is the square root of the quantity

$$(x–X)^2 + (y–Y)^2 + (z–Z)^2.$$

In Special Relativity, there is an analogous quantity, called the *interval* between events (x,t) and (X,T); it is

$$(x–X)^2 – (t–T)^2.$$

Note the minus sign: time is special. That's where your friend H.G. Wells went wrong. Time is another dimension, but it's not like the spatial dimensions. Though it can get mixed up with them, to some extent, as I'll explain in a moment. At any rate, the main point to understand is that along the lines of 45° slope where $(x–X)^2 = (t–T)^2$, so $x–X = t–T$ or $x–X = T–t$, the interval is *zero*. Those 45° lines are called *null curves*.'

'I see that. I have studied the geometry of Monsieur Descartes. But what does this "interval" represent?'

I told him that the interval is related to the apparent rate of passage of time for a moving observer. The faster an object moves, the slower time on it appears to pass. This effect is called *time dilation*. As you approach a null curve – that is, travel closer and closer to the speed of light – the passage of time that you experience slows down towards zero. If you could travel *at* the speed of light, time would be frozen. No time passes on a photon.

'It seems to me that time is somewhat mutable in this theory,' said the Time Traveller thoughtfully.

'That's true. In fact, in 1911 Paul Langevin pointed out a curious feature of Special Relativity, known as the *twin paradox*. Suppose that (Figure 27) two twins, Rosencrantz and Guildenstern, are born on Earth. Rosencrantz stays there all his life, while Guildenstern travels away at nearly light-speed, and then turns round and comes home again at the same

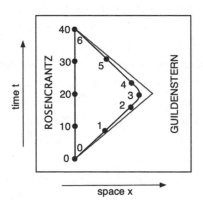

FIG 27 The twin paradox.

speed. Because of time dilation, only six years (say) have passed in Guildenstern's frame of reference, whereas 40 years have passed in Rosencrantz's frame.'

'But surely,' said the Time traveller, 'the circumstances are perfectly symmetric. In Guildenstern's frame of reference, it is Rosencrantz who seems to travel at nearly the speed of light. So by the same argument, it is Rosencratz who ages less. And that is absurd.'

'That's why people think it's a paradox. But it isn't. It only seems paradoxical if you don't actually look at a space-time diagram, because then you may think that it doesn't matter which twin is used as the "fixed" frame. But Guildenstern's motion involves acceleration (positive and negative), while Rosencrantz's doesn't – and that destroys the apparent symmetry between the two twins. Acceleration is *not* a relative quantity in Einstein's theory. Like I said, "relativity" is a silly name.'

The Time Traveller shook his head slowly – I couldn't decide whether he didn't believe what I was saying, or was overawed by its intellectual depth. 'But it is only a theory, of course,' he said, almost to himself. 'Reality is not like that.'

'"Theory" has two meanings,' I said. 'One of them should really be "hypothesis", but that sounds a bit pretentious. That means an idea put up for discussion and experiment. "Only a theory" applies fine to that one. But there's a second meaning: "a body of concepts and results that has survived a long series of stringent experimental tests designed to reveal any flaws." You can't legitimately dismiss anything like *that* with the word "only". "Only an idea that has survived centuries of attempts to shoot it down…" No, that doesn't really work, does it?'

'Well, be that as it may, the effect was tested back in the late twentieth century by transporting atomic clocks around the Earth on jumbo jets.'

'I understand "clock", but little else in that sentence.'

'The clocks were fantastically accurate, and they were carried round the entire planet in very fast flying-machines. Of course, flying-machines are so slow compared to light that the time difference observed (and predicted) is only the tiniest fraction of a second.'

'Um,' said the Time Traveller. 'Flying-machines?'

'You've got a *time* machine – that's much harder to build. Just take my word for it, my friend.'

'So you are telling me that "the time is out of joint", to continue your Shakespearean motif. *Hamlet,*' he added as an afterthought.

'Precisely. So it ought to be possible to exploit that out-of-jointness to make a time machine.'

'Just as I did.'

'Yes. But without that ivory gizmo of yours, we're going to have to use conventional physics, which means relativity. And to do that, we're going to have to understand Einstein's approach to gravity.'

The Time Traveller gawped at me. 'What does gravity have to do with time travel?'

To be continued…

WEBSITES

H.G. WELLS:

http://en.wikipedia.org/wiki/The_Time_Machine
http://en.wikipedia.org/wiki/H._G._Wells

SPECIAL RELATIVITY:

http://en.wikipedia.org/wiki/Special_relativity
http://en.wikibooks.org/wiki/Special_Relativity

POPULAR DESCRIPTION:

http://www.phys.unsw.edu.au/einsteinlight/

TWIN PARADOX:

http://en.wikipedia.org/wiki/Twin_paradox
http://www.phys.unsw.edu.au/einsteinlight/jw/module4
 twin_paradox.htm

8

Forward to the Future 2: Holes: Black, White, and Worm

The story so far...

The Time Traveller has arrived at the offices of Hawkrose & Penking Heavy Engineering. His time machine is seriously damaged, and a lack of elephants makes any repair impossible. Nevertheless, Hawkrose & Penking may be able to help. He has been told about Special Relativity, in which the speed of light is constant.

Now read on...

THE TIME TRAVELLER gawped at me. 'What does gravity have to do with time travel?'

'Everything. Though I admit it's not obvious. You see, Einstein invented another theory, called General Relativity, which was a synthesis of Newtonian gravitation and Special Relativity. You know what Newton said about gravity?'

'I am a highly educated man, sir. It is a force that moves particles away from the perfect straight line paths that they would otherwise follow. The force exerted by any particle of matter varies inversely as the square of the distance.'

'OK. But let's think geometrically. The paths that particles follow, in the absence of any forces such as gravity, are *geodesics*. That is, they are shortest paths, they minimize the total distance between their end points. In flat Minkowski space-time, the analogous relativistic paths minimize the interval instead. The problem is to incorporate the effects of gravity consistently. Einstein's solution was to think of gravity not as an extra force, but as a distortion of the structure of space-time, which changes the value of the interval.

This variable interval between nearby events is called the *metric* of space-time. The usual image is to say that space-time becomes "curved".'

'Curved round what?'

'It's not curved round anything. It's just intrinsically distorted compared to flat space-time. You might as well ask "flat along what?" about ordinary Euclidean space. The curvature is interpreted physically as the force of gravity, and it causes light-cones to deform. One result is "gravitational lensing", the bending of light by massive objects, which Einstein discovered in 1911 and published in 1915. The effect was first observed during an eclipse of the Sun. More recently it has been discovered that some distant quasars – very powerful and very distant cosmological objects – produce multiple images in telescopes because their light is lensed by an intervening galaxy.'

Figure 28 illustrates this idea by showing a space-like section of space-time (in effect, one taken at a 'fixed' instant of time, but the actual description is more technical because relativistic effects imply that 'fixed instant' makes no sense at different locations) near a star. It takes the form of a curved surface that bends downwards to create a circular valley in which the star sits. This space-time structure is *static*: it remains exactly the same as time passes. Light follows geodesics across the surface, and is 'pulled down' into the hole, because that path provides a short cut. Particles moving in space-time at sub-light speeds behave in the same way. If you look down on this picture from above you see that the particles no longer follow straight lines, but are 'pulled towards' the star, whence the Newtonian picture of a gravitational force.

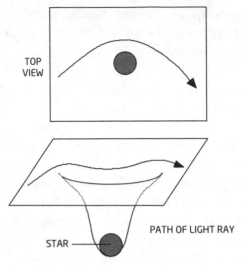

TOP
VIEW

PATH OF LIGHT RAY

STAR

FIG 28 Bending of light by gravity.

'Far from the star,' I told him, 'this space-time is very close indeed to Minkowski space-time; that is, the gravitational effect falls off rapidly and soon becomes negligible. Space-times that look like Minkowski space-time at large distances are said to be *asymptotically flat*. Remember that term: it's important for making time machines. Most of our own Universe is asymptotically flat, because massive bodies such as stars are scattered very thinly.'

The Time Traveller digested this information. 'So I can give space-time any form I wish? That sounds implausibly flexible.'

'No. When setting up a space-time, you can't just bend things any way you like. The metric must obey the *Einstein equations*, which relate the motion of freely moving particles to the degree of distortion away from "flat" Minkowski space-time.'

'I see. There is a connection between the distribution of masses within the space-time, and the structure of the space-time itself. As if matter creates and moulds its own space and time.'

'You're very quick to catch on. It took Einstein years. Anyway, now I can explain how twentieth-century physicists interpreted the phrase "time machine" within the framework of General Relativity.' I could see his interest suddenly increase. He was no longer listening just out of politeness. 'A time machine lets a particle or object return to its own past, so its world-line, a time-like curve, must close into a loop. A time machine is just a *closed time-like curve*, abbreviated to CTC. Instead of asking "is time travel possible?" we ask "can CTCs exist?".'

The Time Traveller leaned forward nervously, and his eyes narrowed. '*And can they?*'

'Well, in flat Minkowski space-time, they can't. Forward and backward light-cones – the future and past of an event – never intersect. But they can intersect in other types of space-time. The simplest example takes Minkowksi space-time and rolls it up into a cylinder (Figure 29). Then the time coordinate becomes cyclic.'

'You mean history perpetually repeats itself, as in Hindu mythology?'

'Sort of. *Space-time* repeats; what happens to history depends upon whether you think free will might be in operation. It's a tricky question and one that Einstein's equations don't really address. They just govern the overall coarse structure of space-time.

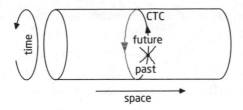

FIG 29 Simple example of a space-time with a CTC.

'Although a cylindrical space-time *looks* curved, actually the corresponding space-time is *not* curved – not in the gravitational sense. When you roll up a sheet of paper into a cylinder, it doesn't *distort*. You can flatten it out again and the paper isn't folded or wrinkled. A creature that was confined purely to the surface would have difficulty noticing that it had been bent, because distances *on* the surface wouldn't have changed – unless it happened to wander all the way round the cylinder. In short the metric – a local property of space-time structure *near* a given event – doesn't change. What changes is the global geometry of space-time, its overall *topology*.'

The Time Traveller sighed. 'Another new word.'

'Topology is a flexible kind of geometry – it studies the properties of shapes that persist when the shape is continuously deformed. Like the presence of holes, say, or knots.'

'Ah. In my day this was called *analysis situs*, the analysis of position. It was very new and known only to a few specialist mathematicians.'

'Well, now it's very old, very respectable, and known to every child before it leaves its gestation-tank. Rolling up Minkowski space-time is an example of a powerful topological trick for building new space-times out of old ones: *cut-and-paste*. If you can cut pieces out of known space-times, and glue them together without distorting their metrics, then the result is also a possible space-time.'

'You are speaking metaphorically, of course.'

'Well, until recently I'd have agreed with you. But when Hawkrose & Penking describes itself as a "heavy engineering" company, it really does mean *heavy. Extremely* heavy. But I'm getting ahead of myself.'

'Like me,' he said with a straight face. I laughed, and not just out of politeness: in his position I'd have had trouble producing any joke, however feeble.

'I say "distorting the metric" rather than "bending", for exactly the reason that I say that rolled-up Minkowski space-time is *not* curved. I'm talking about intrinsic curvature, as experienced by a creature that lives in the space-time; not about apparent curvature as seen from outside. Apparent bending of this type is "harmless" – it doesn't actually change the metric. Now, the rolled up version of Minkowski space-time is a very simple way to prove that space-times that obey the Einstein equations *can* possess CTCs – and thus that time travel is not inconsistent with currently known physics. But that doesn't imply that time travel is *possible*.'

'I see that. There is a very important distinction between what is mathematically possible and what is physically feasible.'

He was sharp, I'll hand it to him. 'Yes. A space-time is mathematically possible if it obeys the Einstein equations. It is physically feasible if it can exist, or could be created, as part of our own Universe. Which is where the heavy engineering comes in. Unfortunately for you, there's no reason to suppose that rolled-up Minkowski space-time is physically feasible: certainly it would be hard to refashion the Universe in that form if it wasn't already endowed with cyclic time. The search for space-times that possess CTCs *and* have plausible physics is a search for more plausible topologies. There are many mathematically possible topologies, but – as with the proverbial Irishman giving directions – you can't get to all of them from here.

'However, you can get to some remarkably interesting ones. In classical Newtonian mechanics, there is no limit to the speed of a moving object. Particles can escape from an attracting mass, however strong its gravitational field, by moving faster than the appropriate escape velocity. In an article presented to the Royal Society in 1783, John Michell observed that this idea, combined with that of a finite velocity for light, implies that sufficiently massive objects cannot emit light at all – because the speed of light will be lower than the escape velocity. In 1796 Pierre Simon de Laplace repeated these observations in his *Exposition of the System of the World*. Both of them imagined that the Universe might be littered with huge bodies, bigger than stars, but totally dark.'

'That is a very curious idea indeed.'

'You said it. They were both a century ahead of their time. In 1915 Karl Schwarzschild took the first step towards answering the analogous question within the context of General Relativity, when he solved the Einstein equations for the gravitational field around a massive sphere in a vacuum. His solution behaved very strangely at a critical distance from the centre of the sphere, now called the

Schwarzschild radius. When it was discovered, its mathematical significance seemed to be that space and time lost their identity in Schwarzschild's solution, and became meaningless. However, the Schwarzschild radius for the Sun's mass is 2 km, and for the Earth 1 cm – measured from the centre, so it is buried so deep that they couldn't go there to see if anything interesting occurred. What would happen to a star that was so dense that it lay inside its own Schwarzschild radius? No one knew.

'Then, in 1939, Robert Oppenheimer and Hartland Snyder showed that such a star would collapse under its own gravitational attraction. Indeed a whole portion of space-time would collapse to form a region from which no matter, not even light, could escape. This was the birth of an exciting new physical concept. In 1967 John Archibald Wheeler coined the term *black hole*, and the new concept was christened.'

The development over time of a static black hole – one that doesn't rotate – is shown in Figure 30, in which space is represented as two-dimensional and time runs vertically from bottom to top. An initial radially symmetric distribution of matter (the shaded circle) shrinks to the Schwarzschild radius, and then continues to shrink until, after a finite time, all the mass has collapsed to a single point, the singularity. From outside, all that can be detected is the *event horizon* at the Schwarzschild radius, which separates the region from which light can escape from the region that is forever unobservable from outside. Inside the event horizon lurks the black hole.

Figure 30a is the sequence of events seen by a hypothetical observer on the surface of the star, and the time coordinate t is the one experienced by such an observer. If you were to

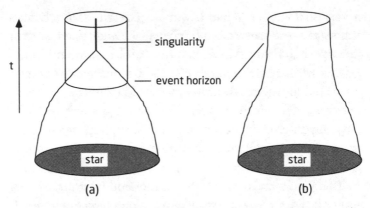

FIG 30 Formation of a black hole as seen by (a) an observer at the surface of the collapsing mass and (b) an external observer.

watch the collapse from outside you would see the star shrinking, towards the Schwarzschild radius, but you'd never see it get there. As it shrinks, its speed of collapse as seen from outside approaches that of light, and relativistic time-dilation implies that the entire collapse takes infinitely long when seen by an outside observer, as in Figure 30b. However, you'd see the light emitted by the star shifting deeper and deeper into the red end of the spectrum. Inside a black hole, the roles of space and time are reversed. Just as time inexorably increases in the outside world, so space inexorably decreases inside a black hole.

'That's where the scope for engineering comes in,' I said. 'Hawkrose & Penking have developed a whole battery of techniques, from quantum foam enlargement to improbability calculus. Because the space-time topology of a black hole is asymptotically flat – like Minkowski space-time at large distances – it can be cut-and-pasted into the space-time of any Universe that has reasonably large asymptotically flat regions, such as our own. This makes black hole topology physically plausible in our Universe. Indeed, the scenario of gravitational collapse makes it even more plausible: you just have to start with a big enough concentration of matter, such as a neutron star or the centre of a galaxy. That's what I mean by heavy engineering. The technology of the thirty-first century can *build* black holes. We use matter-processors – modified neutron stars mostly, with gravitational traps and heavy-duty laser-compressors.

'However, a static black hole doesn't possess CTCs. The next step is to notice that Einstein's equations are time-reversible: to every solution there corresponds another that is just the same, except that time runs backwards. The time-reversal of a black hole is a *white hole*, and it looks like Figure 30 turned upside down. An ordinary event horizon is a barrier from which no particle can escape; a time-reversed horizon is one into which no particle can fall, but from which particles may from time to time be emitted. So, seen from the outside, a white hole would appear as the sudden explosion of a star's worth of matter, coming from a time-reversed event horizon.'

'Why should the singularity inside a white hole suddenly decide to spew forth a star, having remained unchanged since the dawn of time?' protested the Time Traveller.

'Good point. It makes sense for an initial concentration of matter to collapse, if it is dense enough, and thus to form a black hole; but the reverse seems to violate causality. It doesn't, of course – but the cause lies outside our own

Universe, so we don't see the result coming. Let's just agree that white holes are a mathematical possibility, and notice that they too are asymptotically flat. So if you knew how to make one, you could glue it neatly into your own Universe. Hawkrose & Penking have just developed an effective method for doing that, based on the uncertainty principle. We use a Heisenberg amplifier to make the position of matter so uncertain that it may well be outside the normal Universe altogether, and then we can switch on a chronokatoptron to persuade everything to happen in reverse time, since the system doesn't know which time-frame it should be in.

'Not only that: we can glue a black hole and a white hole together. We cut them along their event horizons with a cosmotome and sew the edges together with cold dark matter.' I ignored his blank look. 'The result – more accur-ately, a fixed space-like section of it – is shown in Figure 31: a sort of tube. Matter can pass through the tube in one direction only: into the black hole and out of the white. It's a

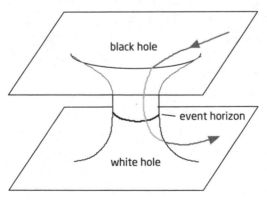

FIG 31 A wormhole.

kind of matter-valve. The passage through the valve is achieved by following a time-like curve, because material particles can indeed traverse it.

'Because the topology of Figure 31 is asymptotically flat at both ends of the tube, both ends can be glued into any asymptotically flat region of any space-time. You could glue one end into our Universe, and the other end into somebody else's. Or you could glue both ends into ours – *anywhere you like* (except near a concentration of matter). Now you've got a *wormhole*.

'Hawkrose & Penking make the best wormholes in the Universe,' I said with pride. 'They're called wormholes because they look like the holes a maggot bores in an apple. Only here the apple is – well, not so much space-time as everything that's *not* space-time.' A wormhole is shown schematically in Figure 32; but you have to remember that the

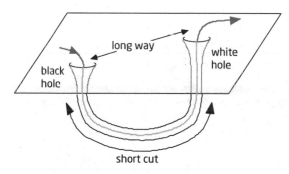

FIG 32 Using a wormhole as a matter-transmitter. (The length of the wormhole is exaggerated in the picture because the picture is drawn in normal space-time. It can actually be very short, even if the ends are far apart in 'normal' space-time, because distance is intrinsic to the space-time in the wormhole.)

distance *through* the wormhole is very short, whereas that between the two openings, across normal space-time, can be as big as you like.

'I see. A wormhole is a short cut through the Universe.'

'Right,' I said. 'But that's *matter-transmission*, not time travel.'

'It nevertheless has some connection with time travel?' the Time Traveller asked urgently, his fingers shaking.

'Well,' I said, 'that'd be *telling...*'

To be continued...

WEBSITES

BLACK HOLES:

http://en.wikipedia.org/wiki/Black_hole
http://hubblesite.org/explore_astronomy/black_holes/
    home.html
http://cosmology.berkeley.edu/Education/BHfaq.html

WHITE HOLES:

http://casa.colorado.edu/~ajsh/schww.html
http://en.wikipedia.org/wiki/White_hole
http://en.wikipedia.org/wiki/White_Hole_(Red_Dwarf_
    episode)

WORMHOLES:

http://en.wikipedia.org/wiki/Wormhole
http://casa.colorado.edu/~ajsh/schww.html
http://webfiles.uci.edu/erodrigo/www/WormholeFAQ.html

9

Forward to the Future 3: Back to the Past, with Interest...

The story so far...

In relativity, 'time machine' means 'closed time-like curve' or CTC. Nothing in the known laws of physics forbids such things. Hawkrose & Penking can take a black hole and a white hole and join them to form a wormhole. But that's matter-transmission, not time travel. Isn't it?

Now read on...

WE STARED AT my picture of a wormhole (Figure 32, previous chapter), hoping for inspiration. 'You do realize,' I said to the Time Traveller, 'that people used to think time travel was a theoretical impossibility, a contradiction in terms?'

'You are referring to the hoary old "grandfather paradox"?'

'Well, he did have an impressive beard, but – oh, sorry, I misunderstood.'

'People raised exactly that objection to my time machine.'

'Yes, the idea goes back to René Barjavel's story *Le Voyageur Imprudent*. You go back in time and kill your grandfather, but because your father isn't born, neither are you, so you *can't* go back to kill him…'

'So you don't so you *are* born, so you do, so…'

'Quite.'

'I only took that objection seriously after I had made my machine,' said the Time Traveller. 'I wondered…people did ask…but no, I quite liked the old gentleman, you see.'

'Don't even think about it,' I said. If you think about the problem using quantum mechanics, you can easily see it doesn't exist.'

'What kind of mechanics?'

'Quantum. New since your day. Quantum mechanics, the underlying physics of matter, is indeterminate. Many events, such as the decay of a radioactive atom, are random. One way to make this indeterminacy mathematically respectable is the "many worlds" interpretation invented by Hugh Everett Jr. This view of the Universe is very familiar to readers of science fiction: our world is just one of an infinite family of "parallel worlds" in which every combination of possibilities occurs. In 1991 David Deutsch noted that, thanks to the many worlds interpretation, quantum mechanics involves no obstacles to "free will". Moreover – another standard science fiction trope – the grandfather paradox ceases to be para-doxical, because grandfather will be (or will have been) killed in a parallel world, not in the original one.'

The Time Traveller digested this for a moment. 'That is a cause for some concern,' he said. 'If I do get back to my home time, how can I tell whether I've accidentally moved to a parallel Universe?'

'Don't worry,' I said. 'According to the many worlds inter-pretation, that's what you're doing every time your constituent atoms choose whether or not to change their quantum state. Which, to be frank, is all the time. You're perpetually switching from whichever Universe you happen to be in at that moment, to a parallel one – one for each possible choice of state.'

'I am not convinced that your attempt to reassure me is terribly comforting, to be frank.'

I hardly heard him. An idea of some sort was brewing in my brain; I could feel my subconscious trying to tell me something. But the Time Traveller was so eager to find a way home that I couldn't get enough peace and quiet for it to pop into conscious thought...

'I think we should forget this quantum mechanics business,' he said, 'and return to a simpler question. Is there a connection between wormholes and time machines?'

Of course! Was *that* what my subconscious was trying to tell me? No, I had a strange feeling it involved money...

'Sure,' I said. 'It was noticed way back in 1988, when Michael Morris, Kip Thorne, and Ulvi Yurtsever realized that they could combine a wormhole with the twin paradox to get a CTC. I'd forgotten it until you asked. The idea is to leave the white end of the wormhole fixed, and to tow the black one away (or zigzag it back and forth) at just below the speed of light.'

Figure 33 shows how this leads to time travel. The white end of the wormhole remains static, and time passes at its normal rate, shown by the numbers. The black end zigzags to and fro at just less than the speed of light; so time-dilation comes into play, and time passes more slowly for an observer moving with that end. Think about world-lines that join the two wormholes through normal space, so that the time experienced by observers at each end are the same: lines joining dots with the same numbers. At first those lines slope less than 45°, so they are not time-like, and it is not possible for material particles to proceed along them. But at some instant, in this case time 3, the line achieves a 45° slope. After this 'time barrier' is crossed, you can travel from the white end of the wormhole to the black through normal

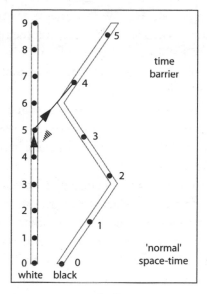

FIG 33 Turning a wormhole into a time machine.

space – following a time-like curve. An example of such a world-line runs from point 5 in the white end of the wormhole to point 4 in the black. Once there, you can return *through* the wormhole, again along a time-like curve; and because this is a short cut you can do so in a very short period of time, effectively travelling instantly from point 4 at the black end to the corresponding point 4 at the white. This is the same place as your starting point, but one year in the *past*! You've travelled in time. By waiting one year, you can close the CTC and end up at the same place and time that you started from. Notice that the corresponding 'ends' of the wormhole are *not* those with the same t-coordinate in

Minkowski space, but those with the same 'elapsed time' for an observer that moves with them, as marked by the figures.

You can make your own wormhole in your own home. Take a plastic bin-liner and cut out the bottom. Fix one end, and imagine the other rushing to and fro at just below light-speed, so that time inside it slows down. When the far end of the bag comes near, walk across to it. When you arrive, that end of the bag will be in your own past. Climb through it, and you'll travel back in time.

If your imagination is vivid enough, that is.

The actual distance you have to travel through ordinary space need not be huge: it depends on how far the right end of the wormhole has to move on each leg of its zigzag path. In space of more than one dimension it can spiral rather than zigzag, which corresponds to making the black end following a circular orbit at close to light-speed. You could achieve this by setting up a binary pair of black holes, rotating rapidly round a common centre of gravity.

'The further into the future your starting point is, the further back in time you can travel from that point,' I told the Time Traveller.

'Wonderful! I can wait several years if necessary.'

'Ah,' I said. 'There's a nasty snag. You can never travel back past the time barrier, and that occurs some time *after* you build the wormholes. There's no hope of getting back to your home time.' His face fell. So did mine. I'd finally figured out what my subconscious was trying to tell me. It did involve money. But it suffered from the same fault.

'There's another problem, too,' I said. 'Hawkrose & Penking's R&D department is working on it, but all we've got is a laboratory prototype. The question is: can you really *build* one of these devices? Can you really get through the wormhole? We can build the wormhole alright, and move its ends around. That's just a matter of creating intense gravitational fields, our stock-in-trade.

'But the problem that bothers me most is what I call the "catflap effect". When you move a mass through a wormhole, the hole tends to shut on your tail. It turns out that in order to get through without getting your tail trapped you have to travel faster than light, so that's no good.'

'Why?'

'The easiest way to see that is to represent the space-time geometry using a *Penrose map*, invented by the twentieth century mathematical physicist Roger Penrose. When you draw a map of the Earth on a flat sheet of paper you have to distort the coordinates – for example, lines of longitude may become curved. The Penrose map of a space-time also distorts the coordinates; but it is designed so that light-cones don't change – they still run at 45° angles. Figure 34 shows a Penrose map of a wormhole. Any time-like path that starts at the wormhole entrance, such as the wiggly line shown, must run into the future singularity. There's no way to get across to the exit without exceeding the speed of light.'

'Which, you have told me, is impossible,' said the Time Traveller.

'Well, maybe not. We're hoping to get round the difficulty by threading the wormhole with *exotic matter*, exerting enormous negative pressure, like a stretched spring. But in 1991 Matt Visser suggested an alternative geometry for a benign wormhole, and we're going to test it out just as soon as we've

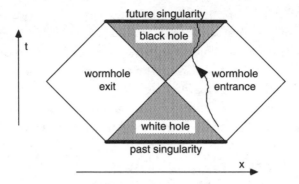

FIG 34 Penrose map of a wormhole.

located a good source of exotic matter. The idea is to cut two identical cubes in space, and paste their corresponding faces together. Then we'll reinforce the edges of the cube with exotic matter.'

'It sounds complicated,' said the Time Traveller.

'Sure is. That's what engineers do. Make complicated things work. However, there's a more old-fashioned method that cuts out the need for exotic matter. And because it doesn't involve *building* a wormhole, there's no time-barrier effect. You can go back to any time you want. Depending on what nature has up her sleeve.' Lots of money, if we struck lucky…

'I don't follow you,' said the Time Traveller, interrupting my beautiful daydreams.

'I'm talking about using a naturally occurring time machine. A *rotating* black hole. Formed when a rotating star collapses gravitationally. The Schwarzschild solution of Einstein's equations corresponds to a *static* black hole, formed

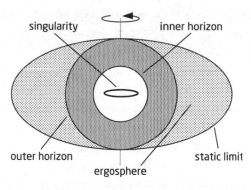

FIG 35 Cross-section of a rotating black hole.

by the collapse of a non-rotating star. In 1962 Roy Kerr solved the equations for a rotating black hole, now known as a *Kerr black hole*. (There are two other kinds of black hole: the Reissner–Nordstrøm black hole, which is static but has electric charge, and the Kerr–Newman black hole which rotates and has electric charge.) It is almost a miracle that an explicit solution exists – and definitely a miracle that Kerr was able to find it. It's extremely complicated and not *at all* obvious. But it has spectacular consequences.

'One is that there is no longer a point singularity inside the black hole. Instead, there is a circular ring singularity, in the plane of rotation (Figure 35). In a static black hole, all matter must fall into the singularity; but in a rotating one, it need not. It can either travel above the equatorial plane, or pass through the ring. The event horizon also becomes more complex; in fact it splits into two. Signals or matter that penetrate the *outer horizon* cannot get back out again; signals or matter emitted by the singularity itself cannot travel past the *inner horizon*. Further out still, but tangent to the outer horizon at the poles, is the *static limit*. Outside this, particles can move at will. Inside it, they must rotate in the same direction as the black hole, although they can still escape by moving radially. Between the static limit and the outer horizon is the *ergosphere*. If you fire a projectile into the ergosphere, and split it into two pieces, one being captured by the black hole and one escaping, then you can extract some of the black hole's rotational energy.

'The most spectacular consequence of all, however, is the Penrose map of a Kerr black hole, shown in Figure 36. The white diamonds represent asymptotically flat regions of space-time – one in our Universe, and several others that need not be. The singularity is shown as a system of broken lines, indicating that it is possible to pass through it (going through the ring). Beyond the singularities lie antigravity Universes in which distances are negative and matter repels other matter. Any body in this region will be flung away from the singularity to infinite distances. Several legal (that is, not exceeding the speed of light) trajectories are shown as curved paths. They lead through the wormhole to any of its alternative exits. The most spectacular feature of all, however, is that this is only part of the full diagram. This repeats indefinitely in the vertical direction, and provides an *infinite number* of possible entrances and exits.

'If we used a rotating black hole instead of a wormhole, and towed its entrances and exits around at nearly light-speed with H&P matter-processing equipment, we'd get a much more practical time machine – one that you could get through without running into the singularity.'

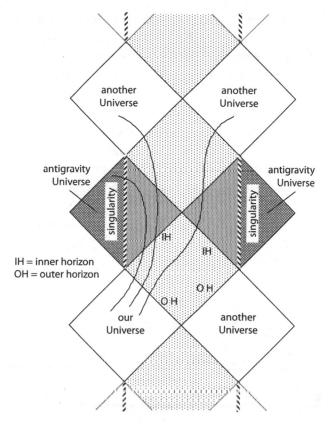

FIG 36 Penrose map of a rotating black hole.

The Time Traveller rubbed his hands together happily. 'Then I shall soon be back in my own time. Come, let us prepare the remains of my machine, to accompany me on the return voyage.'

'Not so fast,' I said. 'Let me check with the computer. Oh, bother. There are no rotating black holes within reach.

There's one under construction, but the union's on strike and it's not been finished yet.' He looked extremely disappointed. Me too. Wait, what had I been watching on the virtual reality hypermedia system the other night? *Got it!* 'I've had an idea, hot off the press. If you don't fancy trying to control Kerr black holes, you can settle for a much simpler kind of singularity: *cosmic string*. This is a static space-time, so that space-like sections remain unchanged as time passes.'

The best way to visualize cosmic string is to use two dimensions of space. Cut out a wedge-shaped sector and paste the edges together (Figure 37a). If you do this with paper you end up with a pointed cone (Figure 37b); but

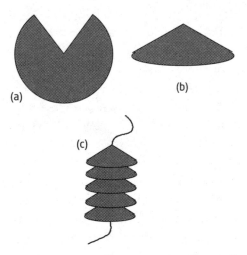

FIG 37 (a) Spatial structure of cosmic string (flattened). (b) Identifying the edges of the missing sector to get a cone. (c) Adding an extra dimension of space.

mathematically you can just identify the corresponding edges without doing any bending. The time coordinate works just as it does in Minkowski space-time (and to get the right shape for light-cones you should identify the edges without making actual cones). If you throw in a third space coordinate and repeat the same construction on every perpendicular cross-section, you get a *line* mass. This is the fully fledged cosmic string. To make a model of one, thread lots of identical cones on a length of – well, string (Figure 37c). Remember, each cone is a constant-time section of the actual space-time.

'I am not sure that I fully comprehend the physical interpretation of a cosmic string as a space-time,' said the Time Traveller.

'Well, basically it's that the cosmic string has a mass, proportional to the angle of the sector that gets cut out. But it doesn't behave like an ordinary mass. Everywhere except the cone point, space-time is locally flat – just like Minkowski space-time. The apparent curvature of a real cone is "harmless". But the cosmic string creates *global* changes in the space-time topology, affecting the large-scale structure of geodesics – particle paths. For instance, matter or light that goes past a cosmic string is gravitationally lensed.'

'Like distant galaxies can bend the light from a quasar?'

'Precisely. Now, in some ways a cosmic string is much like a wormhole, because the mathematical glue lets you "jump across" the sector of Minkowski space-time that is cut out. Way back in 1991, J. Richard Gott exploited this analogy to construct a time machine. More precisely, he showed that the space-time formed by two cosmic strings that whiz past

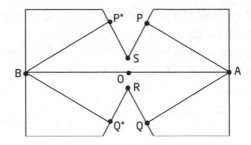

FIG 38 Two cosmic strings, opened flat for clarity.

each other at nearly light-speed contains CTCs. The starting-point is two static strings, symmetrically placed, as in Figure 38, which as usual is a constant-time space-like section.' The time coordinate is suppressed; but if it were added, it would run perpendicular to the page.

'Because of the "gluing", points P and P* are identical, and so are Q and Q*. The figure shows three geodesics joining two points A and B: the horizontal line AB, the line APP*B, and the symmetrically placed line AQQ*B. This demonstrates gravitational lensing by the cosmic strings: an observer at B would see three copies of A, one along each of these three directions.

'Gott calculated that if the two cosmic strings are close enough together, then it takes light longer to traverse the path AB than to traverse the other two. This has an important consequence. If a particle starts from position A but at time T in the past, it can get to B at time T into the future. Call these events A(past) and B(future). If the strings R and S are now made to move, so that S moves rapidly to the right and R rapidly to the left, then A(past) and B(future) become simultaneous in the frame of a stationary observer, thanks to time-dilation.

'So, to construct the required CTC, we make the particle move from A(past) to B(future) passing via PP*; then by symmetry we make it return from B(future) to A(past) via QQ*. Gott's calculations show that provided the cosmic strings travel at close to light-speed, this CTC really does exist – mathematically.'

The Time Traveller scratched his head and grimaced. 'By now I have learned to ask: can such a scenario be realized physically?'

'Well…in 1992 Sean Carroll, Edward Farhi, and Alan Guth proved that there isn't enough available energy in the Universe to *build* a Gott time machine. More precisely, the Universe never contains enough matter to provide such energy from the decay products of stationary particles.'

'It seems yet again that I am trapped forever in my own future.'

'Not necessarily…If we could develop a sufficiently powerful new energy source…but I'm afraid that's not in the works yet. However, I do recall that surveys of the distribution of galaxies in our Universe has revealed that they clump on vast scales, forming structures hundreds of millions of light-years long. This clumpiness is too great to have been caused by gravitational attraction among the known matter.'

'So?'

'One theory is that the clumps were "seeded" by naturally occurring cosmic strings. Provided Hawkrose & Penking's data-banks contain the coordinates of a naturally occurring cosmic string remnant – *and* provided there's a wormhole available to transport you there – we may yet be able to send you home.' And make me a fortune...

'If so, mother Nature has outdone all of the engineering skills of Hawkrose & Penking.'

'Except that you'll need our wormholes to get you to the cosmic string,' I pointed out, as I asked the computer to search for a suitable cosmic string with a nearby wormhole link. A few seconds passed, and then it chirped at me. 'You're in luck,' I said. 'Catch the 3.25 from Lunar Central on the Betelgeuse line, change at epsilon Aurigae to the Ophiuchi hotline, then grab a local commuter to Aldebaran. I'll call a hovercab, pick up your machine, and buy you a ticket.'

'But won't that be expensive?'

'Yes, I said. 'Very. A year's salary. However, there's a way you can repay me.' As I said this I gave the computer new instructions.

'How?' asked the Time Traveller. 'I'd do anything to get back to the end of the nineteenth century.'

The printout whirred into motion. I handed him a sheaf of papers. 'Here is a complete listing of the stock-market prices for major stocks over the entire period 1895–2999. I want you to start a trust fund in my name. Invest one pound in an account with the Bank of England – it's still in existence today and it was in your time too – and use that printout to make sure that the investment grows *very* fast. Understand?'

'Of course. If you can predict the future of the market, your fortune is guaranteed.'

'Exactly. Well, provided we don't get switched to a parallel world. But then, in the past parallel world whose future will become this one, parallel versions of us are probably doing the same thing. There's a lot of convergence to history. I'm willing to risk it. Now, set up a board of trustees to make sure the system keeps working. Take 50% of the profits as an operating fee. Set the trust fund to mature on 27 January 3001 – tomorrow – on presentation of my signature. Here's a specimen signature to put on record.'

'But what if I cheat and keep all the money?' he asked.

'I may just have to come back to the nineteenth century and convince you not to,' I said.

'Oh. Right. Do not worry yourself, I shall do what you ask.'

The hovercab arrived, and he left.

I have a gambling streak to my nature. I'd invested a year's salary getting him back to his own time. But if the gamble pays off…well, let's say I've got an important date tomorrow at the Bank of England.

FEEDBACK

Matt Visser's idea of cutting two identical cubes in space, and pasting their corresponding faces together, bears an uncanny resemblance to a science fiction story that I wrote ten years earlier. Though I didn't do the actual mathematics and physics, you understand, so I'm not claiming to have anticipated him. The story was 'Paradise misplaced', *Analog* **101** no. 3, March 1981, 12-38. The protagonist, Billy the Joat (Jack Of All Trades), is hired to solve a mystery. The archipelago world of Bahamba Bright should have 72,107 islands; now it has 72,106. The small island of Trixydix has disappeared. The passage where Billy discovers *how*, after glimpsing his own face reflected in the ocean, reads like this:

The Joat picked up two chopsticks and laid them side by side on the tablecloth. 'Imagine these are two planes in space,' he said. 'What an interphase transfer plane does is, it kind of slits space along the two planes and glues it all up wrong. It joins the left side of one slit to the right side of the other one, so you get a kind of cross-over effect. Anything going into one plane comes out of the opposite side of the other one. Go into one plane from the left, you come out of the other one on the right, and vice versa. What's more, it doesn't take any time to do it. It just jumps.

'Suppose you set up the machinery needed to create a transfer plane across the base of Trixydix, connected to another one under the ocean somewhere else. Once it's switched on, the planes come into being, and Trixydix seems to end at one of the planes. Above it is just ocean. Now, the plane is perfectly flat, therefore optically flat. The sheared rock - water interface acts as a mirror because it behaves just like a slice of polished rock with water on top. But as soon as the plane ceases to intersect the island, the rest of the interface is water - water, so you don't see anything peculiar. And the water moves freely across the interface, so you can't tell there's a boundary there at all.'

'That's all very well,' said Lindilu, 'but won't you get the top half hovering in mid-ocean over the second plane?'

'Yes, it's got to be more complicated. I'd guess they used a box, with transfer planes for sides. Put a box round Trixydix; put another box round an empty piece of ocean. Cross-connect, and presto! No island.'

WEBSITES

GENERAL:

http://en.wikipedia.org/wiki/Time_travel
http://www.vega.org.uk/video/programme/61

CTC:

http://en.wikipedia.org/wiki/Closed_timelike_curve

COSMIC STRING:

http://en.wikipedia.org/wiki/Cosmic_strings

MANY WORLDS:

http://en.wikipedia.org/wiki/Many-worlds_interpretation

10

Cone with a Twist

You'd think that the geometry of cones
was pretty much old hat. Not so. Glue
two cones together by their bases. Slice
down the middle. If they are just the
right shape, you get a square. Twist one
half through a right angle, and glue the
pieces together again. This is a
sphericon, a delightful mathematical
toy.

THE CONE IS probably most familiar today as an edible container for ice-cream, or – in quantity – as a device for directing traffic away from roadworks. Its past glories lie in higher realms. The geometry of the cone intrigued the ancient Greeks, mainly because of the elegant curves that could be constructed by slicing a cone with a plane. Today the importance of these 'conic sections' – the ellipse, parabola, and hyperbola – rests on their application to celestial mechanics, the movements of planets, comets, and other celestial bodies. The Danish astronomer Tycho Brahe made observations of the planets; the German mathematician, astrologer, and mystic Johannes Kepler calculated that the orbit of Mars must be an ellipse; and the English mathematical physicist Sir Isaac Newton deduced the inverse square law of gravity. The Apollo Moon landings were one consequence.

The Greeks anticipated none of this: they delighted in the intricate geometry of the conic sections for its own inner beauty, and they discovered how to use these curves to solve problems beyond the limited reach of ruler and compasses.

Those problems included trisecting the angle and duplicating the cube (constructing a cube with twice the volume of a given cube – effectively, constructing a line of length $\sqrt[3]{2}$). The new curves could solve these problems because the points where two conic sections meet correspond to solutions of equations of the third and fourth degrees. Ruler and compass can solve only first- and second-degree equations. As it happens, both of these classical problems reduce to solving an equation of the third degree, a fact that is obvious for duplicating the cube, and depends on some simple trigonometry when it comes to trisecting the angle – see Chapter 20. The other famous problem of antiquity, squaring the circle (constructing a square with the same *area* as a given circle) is insoluble even using conic sections – see Chapter 20 again.

The cone itself has generally been of less interest to mathematicians than its planar sections, perhaps because the cone is such a simple shape. Is there anything new left to say about the humble cone? Indeed there is. In May 1999 C.J. Roberts, a reader of the Mathematical Recreations column, wrote to me about a curious shape which he calls a 'sphericon'. He even included two of them – and later sent me a huge box containing several dozen, for reasons I'll explain in a moment.

The sphericon (Figure 39a) is a double-cone – two identical cones joined base to base – but with a twist. Literally. A cone placed on a flat tabletop rolls round and round in circles. A double-cone can roll in a clockwise circle or an anticlockwise one, but it only rolls straight if you bowl it along at speed, or sit it on rails. A sphericon performs a controlled wiggle, which on average is straight. It is easy to make, beautifully simple, and a lot of fun, especially if you make it in

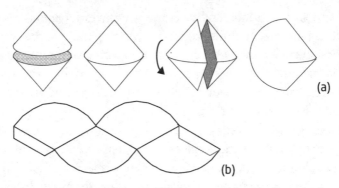

FIG 39 (a) The sphericon. (b) How to make one from card.

quantity. I've never seen such a shape mentioned anywhere before – but who knows what exists out there in the great extelligence?

If you slice a double-cone along a plane that includes both vertices, you get a rhombic cross-section, a parallelogram with all four sides equal. If you use cones of just the right shape, you get a *square* cross-section. Unlike all other rhombuses, the square has an extra symmetry: rotate it through a right angle and it fits back into the same shape. So you can slice such a double-cone down the middle, twist one half through a right angle, and glue the two pieces back together. This is the sphericon. Thanks to the twist, it is not a double-cone, but a much more interesting beast altogether. Two half-double-cones need not make a double-cone!

The sphericon can be made from a single piece of thin card, cut to a shape made from four identical sectors of a circle joined together so that they face in alternate directions (Figure 39b). The main calculation involved in designing this shape is to find the angle between the two straight edges of the sector. Suppose the radius is 1 unit. When the double-cone has square cross-section, the base of each component cone has diameter $\sqrt{2}$, by Pythagoras' theorem. So the circumference of the base is $\pi\sqrt{2}$. The length of the edge of a sector is half that (because you cut the double-cone in half to make a sphericon). The angle of the sector therefore works out as $\pi\sqrt{2}/2$ radians, or $90\sqrt{2}$ degrees, which is roughly 127.28°.

If you cut out the shape shown, you can roll up the sectors into half-cones, and glue the tab to the matching edge. With a little adjustment if necessary, the circular base of the double-cone fits snugly with no gap, and you can tape the join for security if you wish.

The first delight of the sphericon is: *it rolls*! Not only that: it rolls with a wiggle. First one conical sector is in contact with the ground, then the next. So as it moves forward it wiggles alternately to left and right. It is especially fascinating to start it at the top of a slight slope and watch it amble its wobbly way down. When Mr. Roberts's letter arrived, a group of several professional mathematicians spent a pleasant half hour rolling sphericons down a table propped up on books to tilt it.

That letter also hinted at some of the sphericon's fascinating abilities:

> It has one continuous face.
> It will roll on a flat surface.
> One will roll round another, ad infinitum.
> Four will roll round each other in a square block.
> Eight will fit on the surface of one, each one poised to
>   roll, and joined to two others.

This block of nine will roll round another block of nine, ad infinitum.

Intrigued, I asked for more information, and in return was sent an enormous cardboard box, which weighed virtually nothing! This was 'a large lattice of about 50 sphericons' neatly assembled with transparent tape. This lattice, like the atomic lattice of a crystal, repeats indefinitely in three dimensions. I guess I should count myself lucky – or maybe unlucky – that I didn't receive an entire truckload.

One reason why the sphericon has such neat geometric properties is that its four 'edges' – the lines in Figure 39 where the component sectors meet – lie along four of the edges of a regular octahedron. The other four edges of the octahedron correspond to lines that bisect the vertex angles of the sectors, shown dotted. Now the octahedron, in turn, is closely related to the cube – if you put a dot in the middle of each face of a cube and join the dots by straight lines, you get an octahedron. And cubes, of course, stack in a regular manner to form a flat layer or to fill three-dimensional space.

There is, of course, far more to sphericon geometry than this, but it's a good place to start.

Roberts invented the sphericon around 1970. Geometry was always his strong point at school, and he started work as a joiner's apprentice. Not surprisingly, therefore, his first sphericon was carved out of wood. His starting point was the Möbius band, a strip of paper joined end to end with a 180° twist, well known to topologists and schoolchildren. Roberts, however, realized that since paper has a definite thickness, the band's cross-section is really a long, thin rectangle. If you make the cross-section into a square, you can join the ends with a 90° twist instead, getting a solid whose outer surface consists of a single curved face. However, this shape has a hole in the middle: it is a ring. Does there exist a solid that is not a ring whose outside has a single curved face? One day, working on a length of wood with square cross-section, Roberts started thinking about blending one face into the next by planing a curve round the ends. Do this at both ends, eliminate the wood in between, and you get a sphericon.

He made one out of mahogany and gave it to his sister, who has kept it ever since. Then he forgot the topic until 1997, when I gave a series of televised mathematics lectures at Christmas – a regular event in the UK that goes back to Michael Faraday in 1826, though not on TV in those days, of course – and talked about symmetry. At that point his interest was revived, and he wrote to me.

Viewed from the right direction, in line with the middle of a sector, the sphericon looks like a square with one diagonal drawn in (Figure 40a). From another direction it looks like a right-angled isosceles triangle with a semicircle along its longest edge (Figure 40b). If two sphericons are placed next to each other (Figure 40c) then they can roll on each other's surfaces. Figure 40d shows the result after a quarter of a revolution. Figure 40e shows four sphericons arranged so that they can all roll on their neighbours, simultaneously, forever. Eight sphericons can fit around one, all poised so that any one of them can roll on the central one (Figure 41) – though it is not possible to do this so that the surrounding sphericons can also roll on each other. And so on. You can see where that huge box came from.

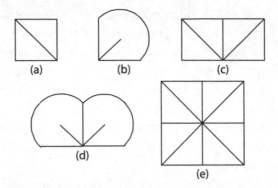

FIG 40 (a) Square cross-section. (b) Triangle plus semicircle. (c) Two sphericons poised to roll. (d) After a quarter turn. (e) Four sphericons poised to roll.

FIG 41 Eight sphericons surrounding one, all poised to roll.

The possible arrangements of sphericons seem endless. I leave you the pleasure of playing with this remarkably simple and extremely clever mathematical toy, and inventing new arrangements for yourselves.

FEEDBACK

Several readers, among them John D. Determan of Alhambra, California, and Cecil Deisch of Warrenville, Illinois, suggested using a cone with a 60° vertex. When sliced in half this has a cross-section that is an equilateral triangle, and two such half-cones can be glued together with a 120° twist. The resulting object rolls, but not far. Deisch came up with an intriguing variant: start with two cones having a 60° vertex, slice them at right angles to a slanting side of the cone, and join them base to base. (The bases are now ellipses.) This object can again be sliced in half to create an equilateral triangle, and two of these can be glued with a twist. David Racusen of Shelbourne, Vermont suggested starting with a cylinder having a square cross-section, and joining two halves with a 90° twist. And Don Bancroft of Brookfield Illlnois sent me a copy of his 1981 US Patent (see Further Reading) describing a rolling device made from two semicircles joined at the middle of their straight sides with a 90° twist. The patent also describes some variants on this idea.

WEBSITES

GENERAL:

http://en.wikipedia.org/wiki/Sphericon

ROBERTS'S WEBSITE:

http://www.pjroberts.com/sphericon/

MOVING '3D' IMAGES:

http://www.interocitors.com/polyhedra/n_icons/index.html

11

What Shape is a Teardrop?

Our senses sometimes deceive us, and here's a case in point. What shape is a teardrop? You won't be surprised to hear that it's not teardrop-shaped. But you may be surprised by how fiendishly complicated it really is.

And the bird on a twig
And the twig on a branch
And the branch on a tree
And the tree in the ground
And the green grass grew all around, all around,
And the green grass grew all around!

THE GUITARIST STRUMMED a final chord, and the singers stopped.

'Thank the Lord for that,' muttered Oliver Gurney.

'If I've said it once I've said it a thousand times –'

'You *have* said it a thousand times,' sighed Deirdre. 'We've all heard you.'

' – the *Potted Dormouse* is not a pub that is improved by folk-singing.'

'It's a pub,' I said, 'with a warm fire. It's raining moggies and doggies outside. I know what gets *my* vote. At any rate, your petition did get rid of the Hammond Organ Night. Though I still don't see why you had to send it to the *Queen*.

The Managing Director of Fosdick's Brewery would have been enough, I'm sure.'

'I believe in going to the very top,' said Oliver. 'Oh God, I bet they'll start on "The Village Pump" next.'

'I *like* "The Village Pump",' said Deirdre. 'I like *all* their songs, they make you see things in a new light.'

'Oh, come now –'

'No, they do. Take that song about the bird on the twig and so forth...it makes you realize how complicated trees are. And how little bits of trees look just like whole trees, only smaller.'

'Self-similarity,' I said. 'Fractal geometry. Big fleas have lesser fleas and so *ad infinitum*. Hence bonsai.'

They're used to me being obscure, and the switch from trees to fleas confused nobody. 'Bonsai?'

'The Japanese art of training little trees to resemble big ones. Wouldn't work unless there was a scale-independent structure.'

'I knew a bloke once did bonsai *mountains*,' said Olly. It took a few seconds for us to twig.

'You mean pet rocks?' enquired Deirdre.

'Suitably fragmented rocks do look a lot like mountains,' I said.

'He didn't just sit a rock in a bowl, you know,' said Olly. 'It's lots of work making proper bonsai mountains. He had all the gear – little hosepipes with spray-action nozzles and fans stuck on special stands to weather them with miniature rain-storms, spark generators for small-scale lightning, lots of tiny mirrors to focus the Sun's rays. Even a tiny snow machine.'

'Really?' Deirdre was interested in gardening and this just about counted.

'Yeah. But he had to stop.'

'Why?'

'The rocks got infested with greenfly. On skis.' Deirdre hit him.

The guitarist stuffed his instrument into its case and propped it against the wall. 'Time for a short break, folks,' he said, and the singers disappeared in the general direction of the bar. Oliver followed them, returning in triumph a few minutes later with two foaming pints and a Blue Moon. He grabbed one pint and Deirdre grabbed the other. Giving me a very strange look, Olly pushed the Blue Moon in my direction.

Look, I've developed a taste for fancy cocktails, OK? I don't have to apologise to anybody. Three-quarters of a measure of vodka, the same of tequila, one measure of blue curaçao, lemonade to taste, all over cracked ice – brilliant. I'd probably move on to a Brooklyn Bomber next, I told him.

Olly grimaced, took a deep draught from his mug, and grinned. 'Beer's better.' He set the glass in front of him. He was about to say something when there was a clear *plink!* sound. We all heard it. Olly looked around for the source, and we heard it again.

'It's your beer,' said Deirdre.

'Beer doesn't go *plink*,' said Olly.

'Yours does. It's drops of rain dripping from the ceiling. There must be a leak in the roof.'

I've never seen Olly move so fast. He grabbed the glass and held it to him like a mother protecting her newborn child from hyenas. 'Dilution,' he explained obliquely. 'D'you think I should have the landlord prosecuted for watering his beer?'

'Olly, it was only *two drops*.'

'It's the principle,' he muttered.

'Well, I have a principle of not annoying perfectly decent landlords. It wasn't done deliberately.'

We watched as the water continued to drip, hitting the table with a splash, throwing tiny beads of spray in all directions. 'I can't see what fascinates you so much,' said Deirdre.

'I'm trying to *see* – no, it all happens too fast. No wonder everybody gets it wrong.'

'Gets what – '

Oliver waved his pudgy hands to silence us. 'Deirdre, you were saying how the folk-songs make you see everyday things in a new light. Raindrops – or teardrops…Let me ask you a question. What shape is a teardrop?'

She thought about it for a moment. 'Teardrop-shaped, of course.'

He passed her a pen and a napkin. 'Draw one for me.' She drew a fat blob rather like a tadpole; round at the head and curving away to a sharp upwards-pointing tail (Figure 42).

Olly looked at it. 'Why do you think it's that shape?'

'Well, that's what they look like. The classic "teardrop" shape.'

'Sure?' Another drip splashed on the table. 'Saw that as it went by, did you?'

'Well, no. It moved too fast. But that's how *everybody* draws them.' Olly nodded, but said nothing. 'You mean, everybody draws them wrong?'

'No comment.'

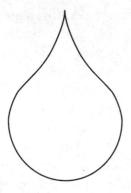

FIG 42 The classic 'teardrop' shape. But is it?

'But when a drip falls off a tap, you get a kind of growing bulge of water hanging down, and then part of it sort of pulls off. So you get a sharp tail formed just before the drip detaches.'

'Draw that too.' She did (Figure 43).

'Hmmm. You reckon the drop keeps its sharp tail as it falls.'

'Yes.'

'But the water hanging from the tap rounds off?'

'Yes. Surface tension.'

'So why doesn't surface tension round off the falling drop's tail too?'

'It gets dragged out behind because the drop's *moving*.'

'Sure?'

Deirdre paused, lips pursed in thought, then shook her head. 'No, it doesn't make sense. The tail would round off as well. Falling teardrops must be roughly spherical. Maybe flattened a bit by air-resistance.'

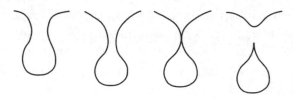

FIG 43 Does a detaching droplet do this?

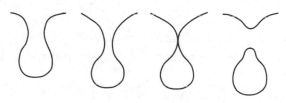

FIG 44 Or this?

Olly nodded. 'Could oscillate, though. So you reckon the picture should really be more like this?' He drew Figure 44.

'I guess. I'm not sure any more.' She looked confused. Isn't it amazing how tenuous our grip on reality can be?

'I've read something about this,' I said. 'The amazing thing, to me, is that the answer wasn't found long ago. Literally miles of library shelves are filled with scientific studies of fluid flow – surely somebody took the trouble to look at the shape of a drop of water? Yet the early literature contains only one correct drawing, made over a century ago by the physicist Lord Rayleigh, and it was *life-sized*.' I paused to draw breath. 'Which means it was so tiny that hardly anybody noticed it.'

'Dead right,' said Olly. 'As a reward, you can buy the next round. The true shape didn't become widely known until

1990, when the applied mathematician Howell Peregrine and colleagues at Bristol University photographed a separating water drop, and discovered that it is far more complicated – but also far more interesting – than you'd ever imagine.' He rapidly sketched out a series of shapes, while I fought my way to the bar. By the time I came back, with two pints and a Harvey Wallbanger (they'd run out of cherry brandy so the Brooklyn Bomber was a non-starter), he'd just about finished (Figure 45).

'That's *weird*,' said Deirdre.

'No, it's just orange juice, vodka, galliano, and a slice of cucumb – '

'Not your drink; the shape of the drop.'

'It's not at all what most people expect,' said Oliver. 'But it's what really happens (Figure 46). It all begins with a bulging droplet hanging from the end of the tap. It develops a waist, which narrows, and appears to be heading towards the classic teardrop shape. But instead of pinching off to form a short, sharp tail, the waist lengthens into a long thin cylindrical thread, with an almost *spherical* drop hanging from its end.'

I picked up the sketch and stared at it. 'I can see why the drop becomes spherical. It's falling so slowly that gravity is

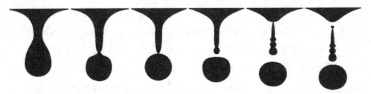

FIG 45 Sequence of changes of shape for a detaching droplet – theory.

FIG 46 Sequence of changes of shape for a detaching droplet – practice.

negligible. So it tries to minimize the energy in its surface tension, and that pulls it into a sphere.'

'Why?'

'Because surface tension is proportional to area, and the sphere has the smallest area for a given volume.' He slapped me heartily on the back. 'But I don't see why that thread forms.'

'Mostly viscosity,' said Olly. 'Stickiness. If the fluid was syrup rather than water, you wouldn't be surprised by a long dangling thread, would you? Water's quite sticky too, though not as sticky as syrup.'

'That's all very well,' said Deirdre. 'But why doesn't the thread go on forever?'

'Instability!' I yelled, startling three old ladies sitting at the next table, playing cribbage. They gave me a sharp look. 'Too long a thread becomes unstable,' I said.

'Exactly,' said Olly, opening a packet of his favourite tripe-and-beetroot crisps. Want some?' he mumbled, waving the bag vaguely in my direction. I shook my head. 'The instability makes the thread start to narrow, right at the point where it meets the sphere, until it develops a sharp point.

At this stage the shape looks like a knitting-needle that is just touching an orange. Then the orange falls off the needle and detaches, pulsating slightly as it falls: the drop has broken off.

'But that's only half the story.' He stuffed more crisps into his mouth and washed them down with a swig of Fosdick's best bitter. 'Now the sharp end of the needle begins to round off, and tiny *waves* travel back up the needle towards its root, making it look like a string of pearls that become tinier and tinier. Finally the hanging thread of water narrows to a sharp point at the top end, and it too detaches. As it falls, its top end rounds off and a very similar series of waves travels along it.'

Deirdre and I both leaned back in our chairs, gazed into space, and then stared at Olly's drawings. 'Astonishing,' said Deirdre. 'I'd never imagined that dripping water could be so *busy.*'

'No,' I said. 'Or so *singular* – and that makes it clear to me why nobody had previously studied the problem in any great mathematical detail.'

'Why not?'

'It's too hard. You see, when the drop detaches, there is a *singularity* in the problem – a place where the mathematics becomes very nasty. The singularity is the tip of the "needle".'

'But why is there a singularity at all? Why does the drop detach in such a complex manner?'

Olly leaped in. 'Because in 1994 the physicists Jens Eggers and Todd F. Dupont showed that the scenario is a consequence of the Navier–Stokes equations of fluid motion.

They simulated the equations on a computer, and reproduced Peregrine's scenario.' He looked like a Cheshire cat with two grins. When he noticed that I wasn't as impressed as he'd hoped, his face fell. 'Why the sour look?' he asked me. 'It was a brilliant piece of work.'

'Absolutely,' I said. 'I'd be proud to have done something half as good. But I just don't think it really answers the question. It's *reassuring* that the Navier–Stokes equations really do predict the correct scenario, but that of itself doesn't help me understand it. There's a big difference between crunching numbers and getting your brain around what the answers *mean*.'

Olly scratched his chin. 'You're talking about the philosophy of explanation again, aren't you?'

'I'm talking about what kind of explanation makes me feel I've understood something. You can dress that up as philosophy, I guess. It certainly isn't science or mathematics as such – it's about how we *understand* science and mathematics.

'The kind of explanation that I'd like to see would be a simple train of logical thought that treats the shape in its own right and convinces me that it has to occur. I'm not sure anybody's yet got an explanation for the falling drop that fits the bill exactly, but I've remembered some work of X.D. Shi and others at the University of Chicago, which is headed in that direction. The main conceptual idea, which was already present in Peregrine's work, is a particular kind of solution to the equations of fluid flow called a *similarity solution*.'

'What's that when it's at home?'

'It's a solution with a certain kind of symmetry, which makes it mathematically tractable. It's temporally self-similar – it repeats its structure on smaller scales at different times. That's why, once the neck of the thread starts to narrow, it *keeps going*, getting narrower and narrower until it forms a point singularity.'

'I don't follow you,' said Olly.

'Not surprising, I'm missing out a lot of mathematical detail. But the idea of a similarity solution explains the shape of the singularity, assuming that a similarity solution exists. That's where the missing technique comes into – '

'Hey,' Deirdre interrupted. 'I've just realized that there's an absolutely classic photograph which shows the singularity perfectly. Only it's milk, not water, and it isn't dripping downwards.'

'Sorry?'

'D'Arcy Thompson's *On Growth and Form*, published in 1942. The first volume has a famous frontispiece, milk splashing into a dish. The splash is shaped like a crown.' See Figure 47.

'Oh, right,' said Olly. 'The photo was taken by Harold Edgerton of Massachusetts Institute of Technology. But it doesn't look like my drawings.'

'Yes it does. Each "spike" in the crown is like a small blob on the end of tube, and the tubes narrow down to sharp points where they meet the blobs.'

'Peregrine's paper did point out that the whole complex series of events is *universal*,' I said. You always see the exact same sequence of shapes when drops detach – in fluids with the appropriate viscosity.'

Oliver decided to test the viscosity of his beer. It slid down very easily, not at all like syrup. 'Did I tell you about the time I invented a tailored bacterium that turned oil into treacle?'

FIG 47 Harold Edgerton's famous milk splash. Each 'spike' in the crown looks like the 'needle-and-orange' picture, the third frame in figure 45.

he asked. 'And very nearly destroyed the entire North Sea oilfiel – '

'Yes, a hundred times,' said Deirdre. 'You saved the day by inventing a tailored yeast to ferment the treacle into alcohol, and created a North Sea beerfield.'[5]

'Long since run dry,' he said sadly.

'Speaking of treacle,' I put in, 'Shi's group took the idea of a similarity solution further, and asked how the shape of the

---

[5] See my science fiction story 'The treacle well', *Analog* 103 no. 10, Sept. 1983, 40–58.

detaching drop depends on the fluid's viscosity. They performed lots of experiments, using mixtures of water and glycerol to get different viscosities. They also carried out computer simulations and developed the theoretical approach via similarity solutions. What they discovered is that for more viscous fluids, a *second* narrowing of the thread occurs before the singularity forms and the drop detaches.'

'You mean you get something more like an orange suspended by a length of string from the tip of a knitting needle?' asked Deirdre.

'Precisely. And now, thanks to the self-similarity of the process –'

She was ahead of me. 'At higher viscosities still, there is a *third* narrowing – an orange suspended by a length of cotton from a length of string from the tip of a knitting needle. And

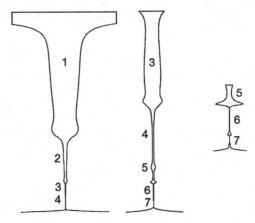

FIG 48 Successive narrowings in a drop of viscous fluid as calculated by X.D. Shi. (Left) The first four narrowings. (Middle) Blow-up of the lower part showing three more narrowings. (Right) Blow-up of the fifth, sixth and seventh narrowings.

as the viscosity goes up, so the number of successive narrow-ings increases without limit. Right?'

'Exactly. At least, provided we ignore the limit imposed by the atomic structure of matter.' See Figure 48.

'Amazing,' said Oliver.

'Never take things for granted,' I said. 'It's the simple questions that have the most surprising answers. But someone has to *ask* the question, not just assume the answer is what everybody would expect.'

'I've got a simple question for you,' said Deirdre.

'What?'

'Do you want another drink? It's my round.'

Olly and I looked at her, then at each other. 'Some simple questions *do* have the answer everybody would expect,' we said in unison.

FEEDBACK

Bonsai mountains feature in Terry Pratchett's humorous fantasy novel *Thief of Time*, the 26th in his Discworld series, which now numbers 32 plus 4 juveniles and numerous spin-offs. Lu-Tze, a history monk, grows them as a hobby. It takes time - but a history monk has all the time in the world, having access to the ancient technology of the procrastin-ator, which can wind and unwind time.

WEBSITES

SPLASHES:

http://courses.ncssm.edu/hsi/splashes/animations.htm
http://en.wikipedia.org/wiki/Harold_E._Edgerton
http://mit.edu/6.933/www/Fall2000/edgerton/www/prewar.
  html
To appreciate how widespread the conventional idea of a
'teardrop' shape is, look the word up on Google™ Images.

12

The Interrogator's Fallacy

More and more, mathematics is tangling with the law. No, they're not making quadratic equations illegal - sorry! DNA evidence has brought probability theory into the courts, thereby opening the proverbial can of worms. So what are the courts doing? Trying to throw the mathematics out again.

MATHEMATICS IS INVADING the courtrooms.

Juries used to be instructed to convict the accused of a crime provided they were sure 'beyond reasonable doubt' of their guilt. This instruction is somewhat qualitative: it all depends upon what each juror considers to be reasonable. A future civilization might attempt to quantify guilt by adopting a common science fiction scenario in which the jury is replaced by a Court Computer. The computer weighs the evidence, calculates a probability of guilt, and terminates the trial when that probability becomes sufficiently close to 1 (which signifies absolute certainty, an ideal seldom attained). But today's civilization does not have Court Computers, so juries are being forced to grapple with probability theory. One reason is the increasing use of DNA evidence. The science of DNA profiling is relatively new, so the interpretation of DNA evidence relies upon assessing probabilities. Similar problems could have arisen when conventional fingerprinting was first introduced, but lawyers were presumably less sophisticated in those days: at any rate, fingerprint evidence is seldom contested on probabilistic grounds. Though even that looks set to change, as more lawyers start to find reasons (sound or not) to dispute the reliability of fingerprints.

In 1995 Robert Matthews – whose work on the 'Anthropo-murphic Principle' was featured in *Math Hysteria* – pointed out that an even longer-standing source of evidence in court cases *ought* to be analysed using probability theory. Namely: confessions. One of Matthews's most surprising conclusions is that there are circumstances under which the existence of a confession adds weight to the view that the accused is inno-cent rather than guilty. He calls this discovery 'The Interro-gator's Fallacy'.

To Tomás de Torquemada, the first Spanish Grand Inquisi-tor, a confession was complete proof of guilt – even if that confession was extracted under duress, as it generally was. Indeed Torquemada authorized the use of torture to obtain evidence, and is estimated to be responsible for some two thousand people being burnt at the stake on the basis of forced confessions. Modern legal practice is generally scep-tical about confessions *known* to have been obtained under duress, but in the mid-1990s there was a series of high-profile terrorism convictions in the UK, that hinged upon confes-sional evidence. The convictions were overturned on appeal because of doubts that the confessions were genuine. Matthews's ideas offer a general reason for distrusting confes-sions in terrorist cases unless they are supported by appro-priate corroborative evidence.

The main mathematical idea required is that of condi-tional probability. This tells us how likely certain events are, given that other events have happened. Human intuition for probabilities is notoriously poor – for instance, we can be unduly impressed by 'coincidences' even when there are prosaic explanations. Things are even worse when it comes to conditional probabilities. Here's a famous case in point.

Mr. and Mrs. Smith tell you that they have two children, and one of them is a girl. They don't say whether the other is a boy or a girl, and for all we know, either is possible. *Given this information*, what is the probability that the other child is a girl? You may assume that at birth, boys and girls are equally likely, each with a probability of ½, and boys or girls occur independently each time. These assumptions are not entirely true, but they're close enough and avoid complications which distract from the reasoning without greatly changing the result.

The reflex response is that the other child is either a boy or a girl, equally likely, so the probability that the child is a girl is 1/2. However, there are four possible gender distributions: BB, BG, GB, GG – where B and G denote 'boy' and 'girl', and the order is that in which the children were born. Each combination is equally likely, and so has probability 1/4. In exactly three cases, BG, GB, GG, the family includes a girl; in just one of those, GG, the other child is also a girl. So in fact the probability of two girls, given that there is at least one girl, is 1/3.

On the other hand, suppose that instead the Smiths tell you that their *elder* child is a girl. What is the probability that the youngest is a girl too? This time the possible gender distributions are BG and GG, and the younger is a girl only for GG, so the probability becomes 1/2. This conclusion seems unreasonable to many, but with the stated assumptions the calculations are correct. They puzzle us because we don't have a good feel for the paradoxical features of conditional probabilities. The two stories of the Smiths' children show that conditional probabilities involve specifying a *context*. The choice of context can have a strong effect on the computed probability. But because the context is usually implicit rather than explicit, we don't pay enough attention to it, and can easily be misled.

Go back to Chapter 1 and look at Figure 2, which lists all 36 ways to roll two dice – each pairing equally likely. Given that at least one die shows a 6, what is the probability that they both do? There are 11 pairs with a 6 somewhere, all equally likely, and exactly one of those has two 6's. So the conditional probability here is 1/11. Now ask a similar question, but with the condition changed to 'the white die shows a 6'. Now there are only 6 pairs that satisfy the condition, so the conditional probability becomes 1/6. The situation for dice is analogous to that for children.

To see how subtle such issues are, suppose that you already know that Mr. and Mrs. Smith have two children, but have no idea of their sexes. One day you see them in their garden (Figure 49). One child, visible, is a girl. The other is partially hidden by the dog, and the sex is uncertain. What is the probability that the Smiths have two girls? You could argue that the question is just like the first scenario above, giving a probability of 1/3. Or you could argue that the information presented to you is 'the child not playing with the dog is a girl', like the second scenario in that it distinguishes one child from the other, so the answer is 1/2. Mr. and Mrs. Smith, who know that the child playing with the dog is young William, would say that the probability of two girls is 0. So who is right?

FIG 49 The Smiths' children and their dog. What is the probability that the child hidden by the dog is a girl?

The answer depends upon a choice of context. Probabilities are about models of reality, not about reality itself. Have you sampled randomly from situations in which there are (in principle) many different families in which either child, randomly, plays with the dog? Or from families in which only *one* child – always the same one – ever plays with the dog? Or are you looking only at a specific family, in which case probabilities are the wrong model altogether?

The interpretation of statistical data requires an understanding of the mathematics of probability *and* the context in which it is being applied. Throughout the ages lawyers have shamelessly abused jurors' lack of mathematical sophistication, either to obtain convictions of innocent people or to acquit the guilty. One example – arising in the context of DNA profiling – is the 'Prosecutor's Fallacy'. I'd like to say that the courts understand this one now, and mostly they do. However, the tragic case of Sally Clark, a solicitor wrongly convicted of murdering her own children, suggests that there is still some way to go. The trial was in 1999, and involved neither DNA profiling nor the Prosecutor's Fallacy: see Websites for the details.

Back to DNA profiling (or DNA fingerprinting, or genetic fingerprinting). First, some background; then we'll see what the fallacy is.

The idea of DNA profiling was invented in 1985 by Alec Jeffreys of the University of Leicester, and centres around so-called VNTR (Variable Number of Tandem Repeat) regions in the human genome. In each such region a particular DNA sequence is repeated many times. VNTR sequences vary greatly between individuals, and are widely believed to identify them uniquely. In 'multi-locus probes', standard techniques from molecular biology are used to look for matches between several different VNTR regions in two samples of DNA: one related to the crime, the other taken from the suspect. Sufficiently many matches should provide overwhelming statistical evidence that both samples came from the same person.

The Prosecutor's Fallacy is based on a confusion of two different probabilities. The 'match probability' answers the question 'What is the probability that an individual's DNA will match the crime sample, given that they are innocent?' However, the question that should concern the court is 'What is the probability that the suspect is innocent, given a DNA match?' Conditional probabilities usually change when the order of the statements is swapped, so the two questions can have wildly different answers. Again, the source of the difference is contextual. In the first case, the individual is conceptually being placed in a population chosen for scientific convenience – say people of the same sex, size, and ethnic grouping. In the second case, they are being placed in a less well-defined but more relevant, and typically smaller, population – those people who might reasonably have committed the crime.

The use of conditional probabilities in such circumstances is governed by a theorem credited to the English probabilist Thomas Bayes. Let A and C be events, with probabilities $P(A)$ and $P(C)$ respectively. Write $P(A|C)$ for the probability that A happens given that C has definitely occurred. Let A & C denote the event 'both A and C have happened'. Then the simplest version of Bayes' theorem tells us that

$$P(A|C) = \frac{P(A\&C)}{P(C)}.$$

This simple case of the theorem is really just a definition of conditional probability, but there is a more general version, which this case illustrates.

For example, in the case of the Smith children, first scenario, we have

C = 'at least one child is a girl'
A = 'the other child is a girl'
$P(C) = 3/4$
$P(A \& C) = 1/4$

because *A* & *C* is also the event 'both children are girls', or GG. Now Bayes' theorem says that the probability that the other child is a girl, given that one of them is a girl, is $(1/4)/(3/4) = 1/3$, the value we arrived at earlier. Similarly, with the second scenario, Bayes' theorem gives the answer $1/2$, also as before.

For the application to confessional evidence, Matthews lets

A = 'the accused is guilty'

C = 'they have confessed'.

As is normal in Bayesian reasoning, he takes *P(A)* to be the 'prior probability' that the accused is guilty – that is, the probability of guilt as assessed from evidence obtained *before* the confession. Let *A'* denote the negation of event *A* (namely 'the accused is innocent'). Then (by a calculation outlined in Box 1) Matthews uses Bayes' theorem to derive the formula

$$P(A|C) = \frac{p}{p + r(1-p)}$$

where to keep the algebra simple we write

$$p = P(A)$$

and

$$r = \frac{P(C|A')}{P(C|A)}$$

which we call the 'confession ratio'. Here $P(C|A')$ is the probability of an innocent person confessing, and $P(C|A)$ is that of a guilty person confessing. The confession ratio is less than 1 if an innocent person is less likely to confess than a guilty one, but it is greater than 1 if an innocent person is more likely to confess than a guilty one.

If the confession is to increase the probability of guilt, then we want $P(A|C)$ to be larger than $P(A)$, which equals p. Therefore we need

$$\frac{p}{p+r(1-p)} > p$$

which some simple algebra boils down to $r < 1$. This inequality has a striking interpretation:

> The existence of a confession increases the probability of guilt
>   if and only if
> An innocent person is less likely to confess than a guilty one.

This actually sounds reasonable, if you think about it. But the implication is less intuitive: sometimes the existence of a confession may *reduce* the probability of guilt. In fact this will occur whenever an innocent person is more likely to confess than a guilty one. But could that ever happen?

In terrorist cases, the answer is 'conceivably, yes'. Psychological profiles indicate that individuals who are more suggestible, or more compliant, or just more easily scared, are more likely to confess under interrogation. These descriptions seldom apply to a hardened terrorist, who will be trained to resist interrogation techniques. An innocent, bewildered person, with no training, subjected to extreme verbal threats, may well confess merely because they are at their wits' end and will say anything to get the interrogation to stop. It is plausible that this is what happened when securing the convictions that were later reversed in UK courts.

Bayesian analysis demonstrates some other counterintuitive features of evidence. For example, suppose that initial evidence of guilt (X) is followed by supplementary evidence of guilt (Y). A jury will almost always assume that the probability of guilt has now gone up. But probabilities of guilt do not just accumulate in this manner. In fact, the new evidence increases the probability of guilt only if:

> The conditional probability of the new evidence, given the old evidence and the accused being guilty
>> exceeds
> The conditional probability of the new evidence, given the old evidence and the accused being innocent.

When the prosecution case depends on a confession, two quite different things may happen. In the first, X is the confession and Y is evidence found as a result of the confession – for example, discovery of the body where the accused said it would be. In this case, an innocent person is unlikely to provide such information, and Bayesian considerations show that the probability of guilt is increased, as we would expect. So corroborative evidence that *depends* upon the confession being genuine increases the likelihood of guilt.

On the other hand, X might be the discovery of the body and Y a subsequent confession. In this case the evidence provided by the body does not depend upon the confession, and so cannot corroborate it. Nevertheless, there is no

'Body-finder's Fallacy' analogous to the Interrogator's Fallacy, because it is hard to argue that an innocent person is more likely to confess than a guilty one just because they know that a body has been discovered.

Of course it would be silly to suggest that every potential juror should take – and pass – a course in Bayesian inference, but it seems entirely feasible that a judge could direct them on simple principles such as those pointed out by Matthews. The Interrogator's Fallacy is not hard to understand. Exactly the same principles apply to DNA profiling, but the Interrogator's Fallacy explains where the reasoning goes haywire in circumstances that are much more intuitive to jurors, and the mathematical point is not obscured by fancy biochemical technology. A brief review of the Interrogator's Fallacy could be an excellent way to discourage lawyers from making fallacious claims about DNA evidence.

BOX 1   MATTHEWS'S FORMULA

By Bayes' theorem

$$P(A\,|\,C) = \frac{P(A\,\&\,C)}{P(C)}$$

and similarly

$$P(C\,|\,A) = \frac{P(A\,\&\,C)}{P(A)}.$$

But C & A = A & C, so we can combine the two equations to get

$$P(A|C) = \frac{P(C|A)P(C)}{P(A)}.$$

Moreover,

$$P(C) = P(C|A)P(A) + P(C|A')P(A')$$

since either A or A' must happen, but not both. Finally, $P(A') = 1 - P(A)$, so if $P(A) = p$ then $P(A') = 1 - p$. Putting all this together, we get the complicated looking formula

$$P(A|C) = \frac{P(A)}{P(A) + \dfrac{P(C|A')}{P(C|A)}P(A')}.$$

Replacing $P(A)$ by p and $P(C|A)/P(C|A')$ by r simplifies this to

$$P(A|C) = \frac{p}{p + r(1-p)}$$

as claimed.

FEEDBACK

I got a lot of mail about the Interrogator's Fallacy, but unfortunately most of it merely confirmed my contention that it is easy to go wrong when considering conditional probabilities. Most readers had trouble not with the main point, the probabilities associated with confessions, but with the preparatory example about the sexes of children. So let me first review the question, which, by the way, is standard both in textbooks on probability theory and in puzzle books, where you will find the same calculation that I carried out. We are told that the Smith family has exactly two children, and that one (or more) of them is a girl. What is the probability that both are girls? We assume boys and girls are equally likely, which is not quite the case in reality.

The big bone of contention was the way I split up the children by considering the order of birth. There are four types of two-child families: BB, BG, GB, GG. Each, I said, is equally likely. The information that at least one is G removes the first case, leaving BG, GB, GG. Of these, only one gives two girls. So the conditional probability that both are girls is 1/3. On the other hand, if we are told 'the elder child is a girl' then the conditional probability that they are both girls is now 1/2. A lot of you disputed these conclusions. Some said that

I shouldn't distinguish BG and GB: there are two cases, B/G and G/G, both equally likely. This is essentially the same mistake that Leibniz made about the odds of getting a double 6 with two dice–see Chapter 1. Instead of arguing theoretically, why don't we just carry out an experiment? Let's toss two coins, and count the proportion of times we get two heads, two tails, or one of each. The coins simulate the sexes, with the right probabilities (1/2 each). Now, if those of you who think BG should not be distinguished from GB are right, then each of these cases should occur about one third of the time. OK, now you go away and do 100 tosses. If I'm right, you should get about 25 cases of two heads, 25 of two tails, and 50 of both. If you're right, you should get about 33 of each.

If you're lazy, like me, you can simulate the tosses on a computer with a random number generator. I did it for one million simulated throws, and here's what I got:

Two heads: 250,025

Two tails: 250,719

One of each: 499,256.

But don't take my word for it: try it for yourself.

The other main argument was that whether or not we know that one child is G, then the other is equally likely to be B or G. That's an interesting argument, and it's instructive to see why it is wrong. Basically, the point is that when both children are girls, there is no unique notion of '*the* other'. It only becomes unique if I specify *which* girl I am thinking about - for example, 'the elder' - which is exactly what makes the two cases different. This destroys the assumed symmetry between Bs and Gs, and changes the conditional probabilities.

In fact, if you think about it, the statement 'the elder child is a girl' conveys *more* information than 'at least one child is a girl'. (The first implies the second, but the second need not imply the first.) So it really ought not to be a surprise that the associated conditional probabilities are different.

Let me also report a development in the legal world, which happened after I wrote the original column. It suggests that the legal profession is not terribly numerate, and believes that juries are even less numerate. In a highly publicized rape case in the UK, a statistician serving as an expert witness explained Bayes' theorem to the jury, in non-technical language, and the accused was found guilty. The defence lawyers then appealed the case on the grounds that jurors who did not wish to use Bayes' theorem were not given an alternative. The appeal failed, but the judges at the Court of Appeal went on record with the view that introducing Bayes' theorem, or anything similar, into a criminal trial 'plunges the Jury into inappropriate and unnecessary realms of theory and complexity, deflecting them from their proper task'. A further appeal was ruled out, leaving the legal status of Bayes' theorem in limbo.

While it is true that juries can be bamboozled by fancy mathematics, the decision has not stopped lawyers doing this, and occasionally high-profile cases still hinge on misuses of probability theory. But now juries seem to have been deprived of entirely sensible mathematical principles that could help them detect such abuses, on the grounds that it's all too difficult for the poor dears.

WEBSITES

GENERAL:

http://ourworld.compuserve.com/homepages/rajm/interro.htm

DNA PROFILING:

http://en.wikipedia.org/wiki/Genetic_fingerprinting
http://en.wikipedia.org/wiki/Prosecutor%27s_fallacy
http://www.dcs.qmul.ac.uk/researchgp/spotlight/legal.html

THE SALLY CLARK CASE:

http://www.sallyclark.org.uk/
http://en.wikipedia.org/wiki/Sally_Clark

13

Cows in the Maze

At last, the cows! To find them, though, you have to solve a maze. Not the usual sort of maze, with hedges and dead ends and suchlike: a logical maze. You'll need two pencils, and the route through the maze depends on which you choose. As an incentive, there's a cow at the end.

MAZES FEATURE FREQUENTLY in recreational mathematics. They are more common in serious mathematics than you might imagine, too, because any mathematical investigation in effect requires you to find a path through a logical maze of statements, with the path from each statement to the next being a valid logical deduction. 'Where are the cows?', a new kind of maze invented by Robert Abbott of Jupiter, Florida, is both a geometric maze and a logical one. It is taken from his book *Supermazes*.

Long-term aficionados of the Mathematical Games column will remember Abbott as the inventor of the card game Eleusis, discussed by Martin Gardner in 1959 and again in 1977. Its appeal relies on a logical twist: the aim of the game – for all players but one – is not to play according to the rules, but to guess what the rules are. The other player's job is to invent them. Abbott's 'cows' maze is also based on a logical twist, that of self-reference. Self-referential statements cause huge problems for logicians and philosophers – for instance the paradox associated with Epimenides, a Cretan who declared that all Cretans are liars, which reduces to:

THIS STATEMENT IS FALSE.

Well, is it, or isn't it? You're in trouble either way. There are *mutually* referential statements like this, too:

THE NEXT SENTENCE IS TRUE.

THE PREVIOUS SENTENCE IS FALSE.

It's a logical minefield.

One way out is to allow the truth of statements to slide on a continuous scale, with half-truths and three-tenths falsehoods; another is to allow the truth of a statement to vary dynamically. In the Mathematical Recreations column for February 1993 I reported work of Gary Mar and Patrick Grim, who discovered that the dynamic approach leads to logical fractals and chaos. Another approach, however, is simply to wallow in the wonder of self-reference, and that's what we'll do here.

As Abbott writes: 'Obviously self-reference is an important area of study for logicians. But the really important question (well, it's really important from my standpoint) is this: can self-reference be used to bring more confusion to mazes? The answer, I am happy to report, is Yes.'

'Where are the cows?' is shown in Figures 50a and 50b, spread across two pages because it's too big to fit on one. Not only is the text self-referential, but the rules for the maze change according to how you move. The text in the boxes is of three kinds: ordinary (roman), **bold**, and *italic*. (In the book they are black, red, and green, but we don't have colour so I've translated everything. That doesn't affect the abstract structure of the maze.) These font types do matter – for instance in boxes 1 and 2.

In order to thread this maze you need both hands, and it helps to hold a pencil or some other pointer in each to remind

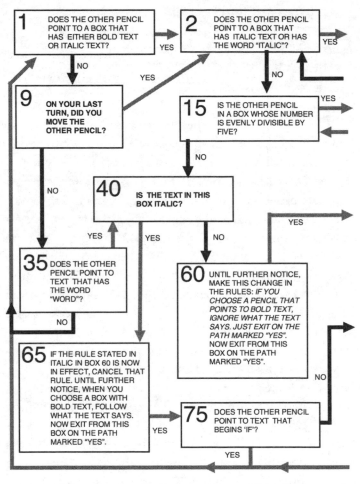

FIG 50a Where are the cows? Start with pencils in boxes 1 and 7, choose a box, and obey the rules. Repeat, and get one pencil to the COW.

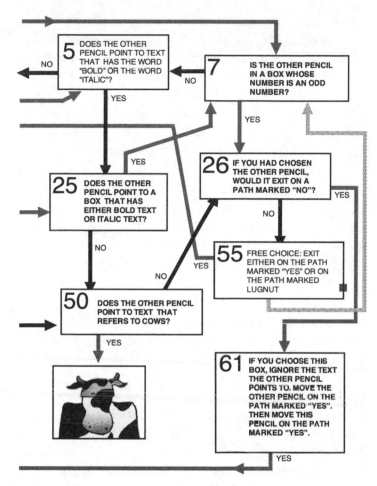

5 DOES THE OTHER PENCIL POINT TO TEXT THAT HAS THE WORD "BOLD" OR THE WORD "ITALIC"?

NO

YES

7 IS THE OTHER PENCIL IN A BOX WHOSE NUMBER IS AN ODD NUMBER?

NO

YES

25 DOES THE OTHER PENCIL POINT TO A BOX THAT HAS EITHER BOLD TEXT OR ITALIC TEXT?

YES

NO

26 IF YOU HAD CHOSEN THE OTHER PENCIL, WOULD IT EXIT ON A PATH MARKED "NO"?

NO

YES

55 FREE CHOICE: EXIT EITHER ON THE PATH MARKED "YES" OR ON THE PATH MARKED LUGNUT

YES

50 DOES THE OTHER PENCIL POINT TO TEXT THAT REFERS TO COWS?

YES

61 IF YOU CHOOSE THIS BOX, IGNORE THE TEXT THE OTHER PENCIL POINTS TO. MOVE THE OTHER PENCIL ON THE PATH MARKED "YES". THEN MOVE THIS PENCIL ON THE PATH MARKED "YES".

YES

FIG 50b Where are the cows? (continued).

you where you are. Or you could use two counters and place them on the boxes.

To start, one pencil points to box 1 and the other to box 7. The numbering on the boxes is not strictly sequential: that's deliberate. Your objective is to make a series of moves so that at least one of the pencils ends up pointing to the box with a picture of a cow, which henceforth is referred to as COW. Abbott labelled this box GOAL and avoided cows, except in box 50, but this extra cow insisted on being allowed into the maze, and does no harm.

To make a single move, *first* choose one of the pencils, *then* follow the instructions in the box to which that pencil points. That's it. No other choices need be made, except when you follow the instructions in box 55. I repeat: *do not follow the instructions in a box until after you have chosen your pencil.* 'Feedback' below shows what can happen if you forget this.

For example, suppose that from the starting position you choose the pencil pointing to box 7. This asks 'Is the other pencil in a box whose number is an odd number?' ('In' here means 'pointing to'.) Now, the other pencil points to box 1, and 1 is odd, so the answer is 'yes'. So you must move the pencil pointing to box 7 along the path labelled YES, which leads it to box 26. The other pencil, pointing to box 1, stays there on this move.

Easy? Just wait. Suppose your next choice is the pencil pointing to box 26. 'If you had chosen the other pencil, would it exit on a path marked 'NO'?' Hmmm. The other pencil was (and still is) pointing to box 1. *If* you'd chosen that, then the question would have been 'Does the other pencil point to a box that has either bold text or italic text?' The 'other pencil' in *this* question is the one that would have still been pointing to box 7, which does contain bold text. So the answer to the question in box 1 is 'yes', and the pencil would have exited along the YES path. All of which means that the answer to the question in box 26 is 'no, it would *not* have exited along the NO path'. So the pencil on box 26 now moves along the NO exit, and ends up pointing to box 55.

Phew.

Most of the boxes ask questions, and your exit path depends upon the answer. Some boxes, however, work differently. Box 61 tells you to move both pencils, and the 'move' is not completed until you have done that. Box 55 has an exit marked LUGNUT instead of the usual 'no'. This does make a difference – for example if your pencils point to 26 and 55 and you choose to move the one on 26.

The really drastic boxes are 60 and 65, which change the rules for threading the maze. Box 60 suspends the usual rule for exiting a box with bold text, replacing it by the rule 'always exit via YES', which I shall call 'rule 60'. Box 65 undoes rule 60 and restores the usual rules. These changes go into effect *only* when you choose the appropriate box – it is not enough merely to have a pencil pointing to one of them. In particular, it is possible to have one pencil pointing to box 60 and the other to box 65. Each box effectively tells you to ignore the other box – but that doesn't cause self-referential problems because you have to choose which one to obey. You don't obey both at once.

Some of the instructions may appear ambiguous, depending on how thinly you believe logical hairs should be split. Box 5 asks whether the other pencil points to text that has the word 'bold' or 'italic'. If, for example, the other pencil points to box 1, the answer is clearly 'yes'; and if it points to box 15, the answer is 'no'. But what if it also points to box 5? Do the quotes round 'bold' mean that the text does not contain the word 'bold', but the word "bold" (with an extra set of quotes)? Abbott's interpretation – which you will need to follow if you are to solve the maze – is that the quotes are irrelevant, and the answer is 'yes' when both pencils are in box 5.

Box 50 asks whether the other pencil points to text refer-ring to cows. A fair question – except that the word 'cows' does not appear in any other box. But of course *both* pencils may point to box 50, in which case the answer is 'yes', so you can exit to COW – unless you want to argue that box 50 does not refer to cows as such: it refers to a *reference* to cows, which is quite a different matter. If that's what you think, you'll never solve the maze, so you should avoid philosophical nitpicking of this kind.

By the way the COW picture (which I've added) is not *text* that refers to cows. But if your pencil points to COW, you've finished the maze anyway, so this issue is irrelevant.

You may by now have convinced yourself that the only possible way to solve the maze is to arrange for both pencils to point to box 50. That would be true in the absence of box 60, which changes the rules. If you can get a pencil to box 50 when rule 60 is in force, then no matter where the other one is, you're done. In fact there is one *other* way to make a legal exit from box 50 along the YES path. Can you find it?

The weirdest situation that could occur has both pencils pointing to box 26. Now the question really is self-referential, and there is no clear way to answer it. So what happens?

Cunningly, Abbott has built his maze so that whenever both pencils point to box 26, rule 60 must be in force, so the text in box 26 is ignored! The same goes if both pencils are pointing to box 61.

What you really ought to do now is have a go, without any further help. If you find that prospect daunting, there are some hints at the end, followed by a complete answer. You may also wish to read Feedback, which warns against common errors when interpreting the rules.

To stop you accidentally reading the hints now, I'm going to ask 'is this really a maze, and if so, in what sense?'.

Traditionally, a maze is a network of *fixed* paths, which may be made from yew bushes planted and cut to shape, or drawn on a sheet of paper. Moreover, you normally expect to move a single object through the maze, not two. With these restrictions there are some general mathematical methods that can be used to thread any maze, notably the 'depth-first search' algorithm, which seeks to explore new territory whenever possible. To understand how it works, first define a 'node' to be any place in the maze where you have a choice between different paths — that is, a place where several paths meet. The procedure for depth-first search is:

1. Begin at the START node.
2. If possible, visit any adjacent node that has not yet been visited, and keep doing this until you can't.
3. In that case, backtrack along your previous path until you find the first node that is adjacent to an unvisited node, and visit that; then go back to stage 2.
4. If you have backtracked along any path, never use it again.

If you keep doing this, you are guaranteed to visit every part of the maze, including the goal – unless the goal is not joined to the start by a path at all, in which case it's a pretty silly maze.

At first sight, this method doesn't apply to 'cows', because of the rule changes, which alter the available pathways, and the choice of which pencil to move. However, that judgement is a little superficial, because 'cows' is equivalent to a more complicated maze of standard type. To begin with, suspend the rule changes from boxes 60 and 65: I'll explain how to deal with them later. First, list all the distinct 'positions', the *pairs* of numbers to which the pencils point. Consider COW to be a number too. For instance (1,7) represents the position when one pencil points to box 1 and the other to box 7. Note that (7,1) represents the same state since it doesn't matter which pencil is which. These number pairs form the nodes of the new maze. Next, list all the possible legal moves – for example we can go from (1,7) to (1,26) or to (2,7) but not to anything else. These moves form the paths connecting nodes together. Now you have set up a conventional maze, and any solution of that translates into a solution of the 'cows' maze. There is just one curious feature: the 'exit' is now any node of the form (COW, ?) or (?, COW), because only *one* pencil has to get to COW to solve the where-are-the-cows? maze.

The rule changes on boxes 60 and 65 are both controlled by rule 60. To deal with that, add an asterisk to positions for which rule 60 is in force. So (1,7) means that one pencil points to box 1, one to box 7, and rule 60 is not in force; whereas (40,50)* means that one pencil points to box 40, one to box

50, and rule 60 is in force. Again, all you have to do is list all the starred *and* unstarred pairs, work out the legal moves, and interpret the results as nodes in a maze and paths between them. Now when rule 60 comes into force you don't alter the maze: you just move into that part of it whose nodes have asterisks. If you wanted to solve 'cows' by brute force, you could set all this up in a computer, do a depth first search, and out would pop the answer.

What if you don't want to resort to brute force? You have several strategies. One is to look for key features of the maze. For example, in order to reach the COW you must have a pencil pointing to box 50, and be in a situation for which the correct exit is YES. As stated earlier, there are three ways to do this. Box 40 has only one exit, YES if rule 60 is in force, NO if not. Another trick is to work backwards from a desired position to see whereabouts you could have come from. And by compiling enough partial paths through the maze you may be able to assemble them into a complete one.

HINTS

If you've tried all that and you're still stuck, here are some hints.

- To reach COW you *must* reach the position (50,50) in which both pencils point to box 50 and rule 60 is not in force. The other two potential ways to finish cannot actually be realized.

- To reach (50,50) you must first get to (35,35). You are then 18 steps away from COW.
- To reach (35,35) you have to get to (61,75) and move the pencil pointing to box 61. Then both pencils can be moved so that they point to box 1. From there it's easy to get to (35,35).
- There are lots of ways to get from the start (1,7) to (61,75). All of them require you to activate the box 60 rule, and then cancel it again at box 65.

ANSWER

In each pair, the underlined number is the pencil that you choose to move. An asterisk shows that rule 60 is in force.

(1,7) (1,26) (2,26) (15,26) (26,40) (26,60) (55,60) (25,55)* (7,55)* (26,55)* (55,61)* (15,61)* (40,61)* (61,65)* (61,75) (1,1) (1,9) (1,35) (9,35) (35,35) (35,40) (35,60) (25,35)* (7,35)*, (26,35)* (35,61)* (1,35)* (9,35)* (2,35)*, (15,35)*, (5,35)*, (5,40)*, (25,40)*, (25,65)* (25,75) (50,75), (50,50), COW.

The part up to (61,75) takes 14 moves, which Abbott conjectures is the minimum. (Has anyone got a proof?) Several alternatives are possible. The rest is the only possible solution, with the slight exception that (5,65)* can be substituted for (25,40)*.

FEEDBACK

'Cows in the maze' was a source of considerable amusement and stimulation. Readers' feedback caused me several moments of panic, with claims of shorter answers, better answers, errors in my (that is, Abbott's) answer, and the like. Several claimed that I was wrong to state that any solution must involve getting to boxes (50,50) with rule 60 not in force. However, when I checked these attempted solutions I found that in every case there was an error.

I'll use the notation of the column, with an underline indicating which pencil is to be moved, and an asterisk showing that rule 60 is in force. One reader's attempt began (1,$\underline{7}$) (1,$\underline{26}$) ($\underline{1}$,55) ($\underline{9}$,15) (35, $\underline{15}$) (35,$\underline{40}$)...However, when moving from position (35,$\underline{15}$) the instruction in box 15 reads 'Is the other pencil in a box whose number is evenly divisible by five?' The answer here is 'yes', and that leads to (35, 5), not (35,40).

A more interesting error occurred in a claimed solution ($\underline{1}$,7) ($\underline{2}$,7) (15,$\underline{7}$) (15,$\underline{26}$) ($\underline{15}$,61) ($\underline{40}$,61) ($\underline{60}$,61) ($\underline{25}$,61)* ($\underline{7}$,61)* ($\underline{26}$,61)* ($\underline{61}$,61)* ($\underline{1}$,61)*, ($\underline{2}$,61)*, ($\underline{15}$,61)* ($\underline{40}$,61)* (65,$\underline{61}$)* ($\underline{75}$,1), (50,1), COW. Its author observed that 'rule 60 is *not* cancelled' as a result of the manoeuvre from (65,$\underline{61}$)* to ($\underline{75}$,1). There is a clear misunderstanding here. If you have arrived at (65,61)* and you choose to move pencil 61, then since rule 60 is in force, you must ignore the bold text - which is all of box 61. This leads you to (65,1) because rule 60 tells you to use the 'yes' exit for the chosen pencil. In order to get to (75,1) you must *obey* the bold text in box 61, which tells you to move both pencils – but you can't do this when rule 60 is in force.

Many misunderstandings arose from confusion about when the rule in box 60 goes into effect. Like all the other instructions, it takes effect only when you *choose* to move the pencil that is currently pointing to that box. It does not take effect as soon as one of the pencils arrives at box 60, because you may not choose that pencil on the next move. Abbott's solution involves a move from (26,60) to (55,60) with rule 60 not in force. Because pencil 26 is chosen, the rule in box 60 is not activated at that time. My correspondent objected on the grounds that the instruction in box 60 contains the word 'now' – but this term is relative. It refers to what you do once you have chosen to move the pencil in box 60: it does not apply until you have made that choice.

WEBSITES

GENERAL:

http://en.wikipedia.org/wiki/Maze

ROBERT ABBOTT'S SITE:

http://www.logicmazes.com/super.html

LOGIC MAZES:

http://www.logicmazes.com/

ONLINE MAZE PUZZLES:

http://www.clickmazes.com/

HISTORY OF MAZES:

http://gwydir.demon.co.uk/jo/maze/

14

Knight's Tours on Rectangles

It's a puzzle at least 1200 years old: move a chess knight around the board to visit every square. Despite a lot of mathematical head-scratching, there's still a lot we don't understand. Even rectangular boards retain a few mysteries. But some of the big questions have recently been solved.

AMONG THE OLD favourites of recreational mathematics are 'knight's tours', in which the chess knight is required to move across boards of various shapes and sizes in such a manner that it visits every square once only. If it can return in one more move to its starting square, the tour is said to be closed. (Recall that the knight moves two squares parallel to a side of the board followed by one more square at right angles.) Figure 51 shows one of the classic knight's tours on a chessboard, discovered by Abraham De Moivre some time before 1800. This one is not closed. For this purpose, the chessboard is merely an 8 × 8 grid of squares, and other shapes were soon investigated.

Knight's tours have a lengthy history. The ninth century Kashmiri poet Rudrata wrote a Sanskrit poem *Kavyalankara*, which encoded a knight's tour on a 4 × 8 board (half a chessboard) in its sequence of syllabic stress patterns. The explicit geometric problem seems to have originated with the English mathematician Brook Taylor around 1700, who asked it for an ordinary 8 × 8 chessboard; the first solutions were sent to Taylor by De Montmort and De Moivre, and appeared in the

FIG 51 De Moivre's tour.

1803 edition of Jacques Ozanam's *Récréations Mathématiques et Physiques*. The first systematic method for finding knight's tours was published by H.C. Warnsdorff in 1823. The problem has since been extended to boards of other shapes, to three-dimensional 'boards', and even to infinite boards.

The literature on knight's tours is extensive but scattered. It includes such classics as *Amusements in Mathematics* by Henry Ernest Dudeney, *Mathematical Recreations and Essays* by Walter William Rouse Ball and Harold Scott MacDonald ('Donald') Coxeter, and *Mathematical Recreations* by Maurice Kraitchik. But in 1991 Allen J. Schwenk (West Michigan University, Kalamazoo) observed that the available modern literature seemed not to contain an answer to an entirely natural question: which rectangular boards support a closed knight's tour? Various sources report that Schwenk's question was solved by Leonhard Euler or Alexandre-Theophile Vandermonde, but fail to indicate either the actual result or its proof. Of the sources listed, Kraitchik comes closest to providing an answer, but assumes that one side of the rectangle has size 7 or less. Rouse Ball deals only with the 8 × 8 case. Dudeney gives several puzzles that reduce to the 8 × 8 case, together with one that requires a tour over the surface of an 8 × 8 × 8 cube.

At any rate, Schwenk took the viewpoint that it is more fun to work out a solution for yourself than to delve into dusty archives. He developed a solution that can easily be explained to students of mathematics, and which illuminates a number of issues in 'discrete mathematics'. Stripped of a few technical details it can be made accessible to almost anyone. Here I shall summarize Schwenk's elegant analysis: for full details see Further Reading.

Mathematically, the knight's tour problem reduces to finding a 'Hamiltonian cycle' in a graph. A graph is a collection of dots (nodes) joined by lines (edges); A Hamiltonian cycle is a closed path that visits each node exactly once. The graph associated to a given chessboard is obtained by placing a node at the centre of each square, and drawing edges between nodes that are one knight's move distant (Figure 52). It is useful to think of the nodes as being black or white, corresponding to the usual pattern of colours on a chessboard.

 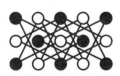

FIG 52 A 3 × 5 chessboard and the corresponding knight's-move graph.

When the knight moves, it hops from a node of one colour to one of the opposite colour, so the nodes must be alternately black and white round any Hamiltonian cycle. This in turn implies that the total number of nodes must be *even*. The 3 × 5 board has 15 nodes, an odd number, so we have proved (without even trying) that no closed knight's tour is possible on the 3 × 5 board. The same goes for any rectangular board of size $m \times n$ where m and n are both odd.

This kind of argument is known in mathematical circles as a parity proof, because it depends on the distinction between odd and even, and its application to knight's tours is well known. Less well known is a more subtle parity proof, discovered by Solomon Golomb and in modified form by Louis Pósa, demonstrating that there is no closed knight's tour on any 4 × n board. Pósa's version introduces a second colouring in which the top and bottom rows of the board are 'red' and the two middle ones are 'blue' (Figure 53); Golomb's proof combines both colourings.

Here I will describe the proof using Pósa's approach. It is no longer true that blue nodes are joined only to red nodes, because some blue nodes are joined to blue ones. However, every red node is joined only to blue nodes. Thus any presumptive Hamiltonian cycle consists of single red nodes

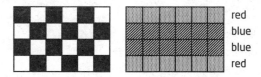

red
blue
blue
red

FIG 53 Golomb's and Pósa's colouring method.

separated by chains of blue nodes. But the numbers of red and blue nodes are the same, so red and blue nodes must alternate round the cycle. But the same is true of black and white nodes, using the more traditional colouring. So by starting at the top left-hand corner, we conclude that all red nodes are black and all blue nodes are white. Since the two colouring schemes are obviously different, this is absurd, so the presumed cycle cannot exist.

We can now state Schwenk's beautiful characterization of those rectangular boards that support knight's tours. An $m \times n$ chessboard (we here take $m \times n$ to avoid duplications) supports a knight's tour *unless*

- m and n are both odd
- $m = 1$, 2, or 4
- $m = 3$ and $n = 4$, 6, or 8.

Let me sketch the proof. We have already disposed of the cases m, n both odd and $m = 4$. It is easy to see that when $m = 1$ or 2 there just isn't room for the knight to get around the board. In fact, the top left-hand node has only one edge connected to it, so no closed cycle can pass through it. The 3×4 case is taken care of by Pósa's argument. For the 3×6 case, observe that removal of two nodes at the top and bottom of column three divides the graph into three disconnected pieces; however removal of two nodes form a Hamiltonian cycle always produces *two* disconnected pieces. The 3×8 case is more complicated and you should either consult Schwenk's article or try it for yourself. (If you can find a *simple* impossibility proof for the 3×8 case, let me know.)

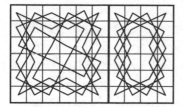 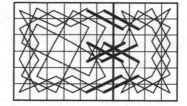

FIG 54 Adding four columns to an existing tour (here the 6 × 6 tour on the left): cross-connect using the thick lines.

That completes the proof of impossibility in the designated cases. It remains to prove that tours exist on all other sizes of board. The key idea now is that a tour on an $m \times n$ rectangle can always be extended to one on an $m \times (n + 4)$ rectangle, provided that certain technical conditions about the existence of certain edges in the tour are satisfied (Figure 54). Moreover, those technical conditions remain valid for the tour on the larger rectangle, so the extension procedure can be continued indefinitely. By symmetry, a tour on an $m \times n$ rectangle can always be extended to one on an $(m + 4) \times n$ rectangle.

So, for example, if we start with a tour on a 5 × 6 rectangle, we know we can also find tours on rectangles of sizes 5 × 10, 5 × 14, 9 × 6 (hence 6 × 9), 9 × 10, 9 × 14, 13 × 6, 13 × 10, 13 × 14, and so on. Each 'initial size' generates a whole family of sizes for which the existence of a knight's tour is guaranteed. The final step is to find enough different initial sizes for this process to generate all the required sizes. It turns out that nine are enough: the boards of size 5 × 6, 5 × 8, 6 × 6, 6 × 7, 7 × 8, 6 × 8, 8 × 8, 3 × 10, and 3 × 12 (Figure 55). Golomb had previously solved the 10 × 3 case. Starting from these, and

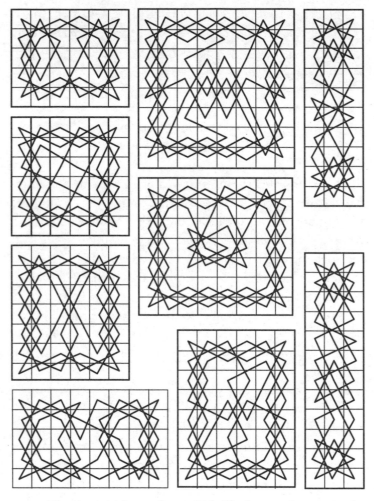

FIG 55 The nine initial cases from which all others can be generated.

the corresponding diagrams rotated through a right angle, and repeatedly adding multiples of four to each side, we can generate tours of all the possible sizes. The proof is complete.

FEEDBACK

Andy Campbell of West Hartford, CT recalled the long-standing problem of a magic knight's tour. This is a closed knight's tour of the 8 × 8 board with the property that, if successive positions of the knight are numbered 1 through 64, the numbers form a magic square. That is, all row sums, column sums, and diagonals are equal. Until recently (and when I wrote the column) the existence of such a tour had neither been proved nor disproved. However, several 'near misses' were known. A square that is magic except for its two diagonals is said to be *semimagic*, and in 1882 E. Francony discovered a semimagic knight's tour (Figure 56). All rows and columns total 260, but the diagonals give 264 and 256.

This 'near miss' stood the test of time, and for a good reason. In 2003, after a computation occupying more than two months of total processing time, it was proved that no such tour exists. The proof was carried out using 'distributed computing', meaning that volunteers could download the software and carry out an assigned part of the task on

2	59	62	7	18	43	46	23
61	6	1	42	63	24	19	44
58	3	60	17	8	45	22	47
53	16	5	64	41	20	25	36
4	57	52	9	32	37	48	21
15	54	13	40	49	28	35	26
12	51	56	31	10	33	38	29
55	14	11	50	39	30	27	34

FIG 56 Knight's tour forming a magic square except for the diagonals.

their own computers in their own time. The software was written by Jean Meyrignac, and Günter Stertenbrink set up a website so that the volunteers could tackle their parts of the problem independently and send in their results. Between them they found 140 different semimagic tours, but after exhausting all of the possibilities, none was magic. See Websites below.

Richard Ulmer, of Denver Colorado, noted that my 6×6 tour is one of the ten tours (out of 9862 in total) to possess 90-degree rotational symmetry (Figure 57). I had remarked that the smallest value of n for which there is a tour on the $3 \times n$ board is $n = 10$. He calculates that there are exactly 16 tours on this board, 176 on the 3×11 board, 1536 on the 3×12, and so on up to a staggering

107,141,489,725,900,544

distinct tours on the 3×42 board. There are eight solutions on the 5×6 board, 44,202 on the 5×8, and 13,311,268 on the 5×10.

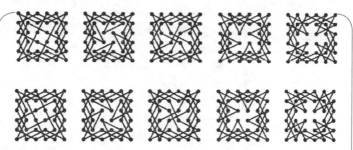

FIG 57 The ten rotationally symmetric tours on a 6 × 6 board.

He also has information about symmetries. For example, no knight's tour can have diagonal flip symmetry. On a rectangle whose two sides are even numbers, no tour is symmetric about a major axis. When the vertical side is odd, a tour with horizontal flip symmetry is still impossible. However, flip-symmetric tours do exist on some boards (Figure 58). Specifically, they can occur when one side is odd and the other is twice an odd number. His current conjecture is that - perhaps with a few small exceptions - flip-symmetric tours exist on all such boards. A proof is currently lacking - so here's another neat problem to get your teeth into.

WEBSITES

GENERAL:

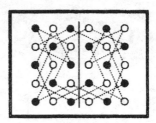

FIG 58 Bilaterally symmetric knight's tour on a 6 × 5 board.

LINKS:

http://www.velucchi.it/mathchess/knight.htm

HISTORY OF MAGIC KNIGHT'S TOURS:

http://www.ktn.freeuk.com/1d.htm

NON-EXISTENCE OF MAGIC KNIGHT'S TOUR ON AN 8 × 8 BOARD:

http://magictour.free.fr/
http://mathworld.wolfram.com/news/2003-08-06/magic
  tours/

MAGIC KNIGHT'S TOUR ON A 12 × 12 BOARD:

http://www.gpj.connectfree.co.uk/gpjh.htm

15

Cat's Cradle Calculus Challenge

All you need is a loop of string, and a friend to help out when two hands really aren't enough. 'Cat's cradle' is just one of a huge range of string figures, found in many cultures. But what's mathematical about that?

THIS CHAPTER IS about a piece of recreational mathematics that – as far as I knew when I first wrote about it – doesn't exist, and mostly still doesn't, but should. What I asked for was a 'calculus' for traditional string figure games such as 'cat's cradle' and its innumerable variants. I'll follow the original column by setting this prospect up as a challenge and describing some of the phenomena that such a calculus should capture. The Feedback section brings the topic up to date and explains to what extent the challenge has been met.

String figures occur in many places, including literature. In Kurt Vonnegut's science fiction novel *Cat's Cradle*, the world as we know it comes to an end when all the seas freeze over into ice-nine, a hypothetical variant of normal ice that is solid at room temperature. Ice-nine is the creation of Dr. Felix Hoenikker, who bequeathes a tiny chip of the substance to his three children Angela, Frank, and Newt. Felix is an inadequate father – which, in the end, is why that chip of ice-nine escapes and freezes the oceans, the rivers, and most living creatures. In a couple of places, Vonnegut alludes to the book's title. The closest that little Newt ever sees his father come to playing a game is when Felix borrows a length of string and makes a cat's cradle out of it. 'He all of a sudden came out of his study and did something he'd never done before,' Newt relates. 'He tried to play with me.' But the attempt was a dismal failure, and much later in the book, Newt explains why:

'For maybe a hundred thousand years or more, grownups have been waving tangles of string in their children's faces…No wonder kids grow up crazy. A cat's cradle is nothing but a bunch of X's between somebody's hands, and little kids look and look and look at all those X's…'

'And?'

'No damn cat, and no damn cradle.'

Vonnegut's story needs a cynic, and Newt fits the bill – but his diagnosis of the cause of Newt's childhood tribulations is probably not widely applicable. String figures, of which cat's cradle is the best-known example, have been popular for centuries in many cultures, and the children enjoy them just as much as the adults. To be sure, you need a bit of imagination to see the alleged cat. The cradle is rather more credible.

The basic cat's cradle game is well known, but not everybody realizes that the complete cat's cradle sequence involves *eight* separate figures. Moreover, innumerable other figures can be constructed in the same general manner, with a simple loop of string held between the fingers of two hands, draped and twisted round them. Although string figures lack explicit mathematical features, they are the kind of thing that should interest any recreational mathematician, with their curious mixture of geometry, topology, and combinatorics. They illustrate the extent to which the topology of a loop of string fails to capture its richer geometric properties, such as shape. To a topologist, all the forms that can be made by twisting and tangling the original loop are, in effect, exactly the same as that loop. But to a geometer, they are not – and the range of possible shapes is beautiful and surprising.

Maybe Newt Hoenikker was a topologist.

I think that it ought to be possible to devise a 'calculus' of cat's cradle shapes, a kind of algebra describing how to get from the initial uninteresting loop to more significant shapes, by making sequences of standard 'moves' of various kinds. The subject known as knot theory – especially that part of it called 'braids' – proceeds in much this manner. Its aim, however, is to capture when two loops are topologically the same, whereas the aim of cat's cradle calculus should be to capture when two topologically equivalent loops are geometrically *different*.

To follow the instructions below, you need a piece of soft, smooth string about three feet (one metre) long, with its ends tied to form a closed loop. The full cat's cradle sequence is shown in Figure 59. It requires two players, Angela and Bill, who take turns to remove the loop of string from each other's hands. First, Angela sets up the cradle (Figure 59a,b). There is one basic movement in the sequence, used at almost every step, and this is the first place it arises. Bill stands on (say) Angela's right. Looking down into the figure, he can see two crossings: he picks these up, one in each hand, and pulls them apart (Figure 59c). Then he draws the strings away from the centre of the figure, over the outside edge, down, inwards,

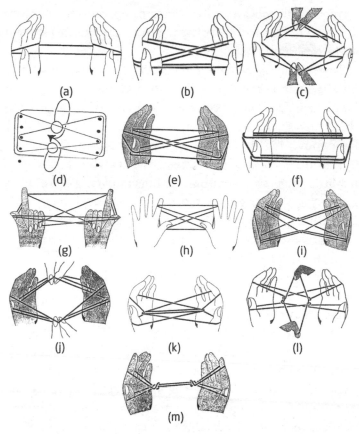

(a)   (b)   (c)

(d)   (e)   (f)

(g)   (h)   (i)

(j)   (k)   (l)

(m)

FIG 59 Cat's cradle.

and back up through the gap in the centre (Figure 59d). As Bill draws his hands apart and separates his thumb and index finger, Angela loosens the loops from her fingers and lets them slip off. Now Bill can take the new figure onto his hands (Figure 59e). This second stage is called the *soldier's bed*.

If Angela now repeats exactly the same moves, starting from this second figure, she creates the third (Figure 59f) known as the *candles*.

To get from the candles to the fourth figure requires a new movement. Bill first draws aside the two inner strings with his little fingers, and passes the thumb and index finger into the centre of the figure from below. This is similar to the basic move, but no crossed strings are carried. Finally Bill opens up his thumb and forefinger, and grips the loops round his little fingers by bending the fingers over. The result is Figure 59g, the *manger*. As a mathematical aside, the manger is just like the cat's cradle, but upside down, so the entire sequence could now be followed in reverse. The traditional route, however, takes unexpected turnings.

From the manger, another repetition of the basic move, also performed upside down (take the crossings from below rather than above), leads – as you might expect – to the soldier's bed upside down (Figure 59h). Traditionally this fifth shape is called the *diamonds*. Yet another repetition of the basic move, this time the usual way up, produces the *cat's eye* (Figure 59i). Picking up slightly differently (Figure 59j) and drawing the hands back *without* swooping back underneath to the centre, leads to the *fish-on-a-dish* (Figure 59k).

The final shape is more elusive. Bill uses his little fingers to separate the central strings (Figure 59l) and then picks up the crossings in the usual manner. Then he turns his thumbs and index fingers inwards and upwards, to get the eighth shape, the *clock* (Figure 59m). I have no idea why the shape has this name, and on this occasion I sympathize somewhat with Newt.

If you use different moves, you can change the order of the sequence – for instance by going straight from the cradle to the candles, or from the soldier's bed to the cat's eye. An effective Cat's Cradle calculus ought to be able to handle all such variations.

The sequence just described is common to many cultures, but the names vary considerably, and include:

- *Cradle:* hearse-cover, water
- *Soldier's bed:* chessboard, mountain cat, church window, fishpond
- *Candles:* chopsticks, clog soles, musical instrument, mirror
- *Manger:* upset cradle
- *Diamonds:* squares
- *Cat's eye:* cow's eyeball, horse's eye, diamonds
- *Fish-in-a-dish:* musical instrument, rice grinder
- *Clock:* considering how little this resembles a clock, it's curious that it alone seems to have no other name.

As an example of the many alternatives that can be formed, I'll give instructions for a shape that can be made by a single player, using a more elaborate series of moves. This figure, *Indian diamonds*, starts in a very similar way to cat's cradle, but not quite (Figure 60). Begin with the standard loop (Figure 60a), then pick up the string that runs across the left palm with the right index finger (Figure 60b), and repeat with the other hand (Figure 60c). Next, slip the loop off your thumbs by bending them in towards each other and gently but steadily pulling your hands apart. Twist your hands so

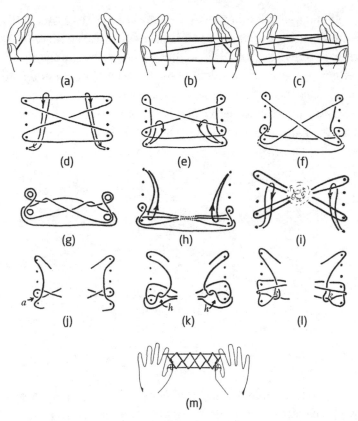

(a)

(b)

(c)

(d)

(e)

(f)

(g)

(h)

(i)

(j)

(k)

(l)

(m)

FIG 60 Indian diamonds.

that your palms face outwards. Pass your thumbs forwards under all the strings, hook them over the little-finger string, and twist your hands back drawing the little-finger string towards you (Figure 60d). This motion is more natural than it sounds, and if you try it, you'll find that the string that you pick up is the 'obvious' one for this method.

Figure 60e shows what the strings now look like, and what to do next. Pass your thumbs over the top of the string immediately in front of them, then underneath the next strings to pick them up with the backs of the thumbs, to get Figure 60f. Next, slip the loops off your little fingers by bending the fingers and pulling your hands gently apart. The result (Figure 60g) is rather tangled, but from here on out it gets simpler. Figure 60h shows the next move: bend your little fingers towards you, turning the hands over if you wish, and bend the fingers over the first string they meet (from the index fingers) and under the next string after that (from the thumbs). Now straighten the little fingers.

At this stage there are two loops on each thumb, and you should free these, just as before. After this the string looks a lot simpler (Figure 60i) except for a tangled knot in the middle, which I won't bother to illustrate since it's irrelevant. Pass the thumb over the two strings that make a loop at the index finger, then under the nearer string of the little-finger loop, and back to where you started from. You may need to twist your hands a bit here.

The string should look like Figure 60j. The next step is unusual. Using the fingers of the right hand, pick up the string at the point marked 'a' and lift it over the left thumb, a fraction of an inch away. Then repeat on the other hand. Be careful to pick up the string *above* the string from the little finger that crosses it. If you've done this correctly, you'll end up with Figure 60k – again with the details of the knotty middle omitted.

Almost there. The final step is easier to do than to describe. Turn your thumbs to point towards each other, pass them through the holes marked 'h' in Figure 60k, below, and bring them up on the near side. Then point your index finger into the holes marked 'k' in Figure 60*l*. Carefully slip the string off the little fingers, and turn your palms smoothly outwards to stretch the string out. You should, after a bit of practice, get Figure 60*m* – Indian diamonds in all their glory.

These two examples have merely scratched the surface of string figures. If you want to know more, take a look at Caroline Jayne's *String Figures and How to Make Them*.

FEEDBACK

Mark A. Sherman, editor of the *Bulletin of the International String Figure Association*, sent me several copies of his journal and its predecessor containing articles that head in the right direction. Among them are a special issue of *Bulletin of the String Figures Association* by Tom Storer and articles in the *Bulletin of the International String Figure Association* by Mark A. Sherman, Joseph D'Antoni, Yukio Shishido, and James R. Murphy. The full references are listed in Further Reading.

The most mathematical response was from Martin Probert, who has posted a series of relevant articles on the Internet, listed below. His results include a method for analysing string figures that resemble each other except for differences in which string overlaps which at a crossing, and some ideas about 'motifs' – common sub-patterns in string figures. There are also a number of new string figures, such as the Jabberwock and Alice in Wonderland, both invented in 2002.

WEBSITES

GENERAL:

http://www.alysion.org/string.htm
http://en.wikipedia.org/wiki/String_game
http://en.wikipedia.org/wiki/Cat%27s_cradle

INTERNATIONAL STRING FIGURE ASSOCIATION:

http://www.isfa.org/

MARTIN PROBERT'S WEBSITE:

http://website.lineone.net/~m.p/sf/menu.html

16

Glass Klein Bottles

Topology is rubber-sheet geometry, but
most mathematicians prefer the traditional tools of blackboard and chalk –
when they're not using a supercomputer.
Alan Bennett has a different approach.
He likes to make things from glass. He
even proves theorems that way.

A DOZEN OR MORE years ago, Alan Bennett, a glassblower from Bedford, became intrigued by the mysterious shapes that arise in topology – Möbius bands, Klein bottles, and the like – and he came across a curious puzzle. A mathematician would have tried to solve it by doing calculations, an artist would have drawn pictures. Alan reached for the materials most familiar to him, and solved it in glass. His series of remarkable glass objects, in effect a research project frozen in glass, became a permanent exhibit at the Science Museum in London.

Recall that topologists study properties of shapes that remain unchanged when those shapes are stretched, twisted, or otherwise distorted – the sole proviso being that the deformation must be continuous, so that the shape is not permanently torn or cut. However, there is one further possibility that I didn't mention in previous discussions of topology because it wasn't relevant then. It is also permissible to cut the shape temporarily, provided it is eventually joined back together again so that points that were originally adjacent across the cut end up adjacent again. This convention – an informal interpretation of the technical concept of 'continuous transformation', not just the *ad hoc* proviso that it may seem – allows mathematicians to treat the shape in its own right and ignore any surrounding space. Topological properties include connectivity – is the shape in one piece, or several? Does it have holes in it? If so, what kinds of holes?

Knots and links are trickier. They have topological properties too, but now the surrounding space is explicitly taken into account when formulating the mathematical concepts. A closed loop with a knot in it is topologically equivalent to a closed loop without a knot – all you have to do is cut the loop, untie the knot, and rejoin the cut. However, the knotted loop sits inside space in a different way from the unknotted loop. There is no way to distort *the entire space* topologically so that the knotted loop becomes unknotted, even when cutting and pasting is allowed – because you must cut and paste *the entire space*, not just the loop.

Topology is a relative newcomer to mathematics. After some early prehistory, it first got off the ground as a subject in its own right about a hundred years ago, when the great French mathematician Henri Poincaré introduced some of the basic algebraic techniques. Its tentacles now extend into every area of modern mathematics, both pure and applied. It has, for example, become indispensable in celestial mechanics, the study of many bodies moving under gravity, where it describes the possible kinds of motion and classifies different kinds of collision.

The most familiar topological shapes appear at first sight to be little more than curious toys, but their implications run deep. There is the Möbius band, which you can make by taking a long strip of paper and gluing its ends together after giving the strip a twist. Throughout this chapter, 'twist' means 'turn through 180°' – sometimes this operation is known as a half-twist. The Möbius band is the simplest surface that has only one side. If two painters tried to paint a Möbius band red on one side and blue on the other, they would eventually run into each other. If they tried the same game on the hollow surface of a sphere, they wouldn't have that problem. The sphere would end up with one red surface – the outside, say – and a blue surface inside. A sphere is two-sided, a Möbius band isn't. Get used to it.

If you give the strip several twists, you get variations on the Möbius band. To a topologist, the important distinction is between an odd number of twists, which leads to a one-sided surface, and an even number, which leads to a two-sided surface. All odd numbers of twists yield surfaces that, intrinsically, are topologically the same as a Möbius band. To see why, just cut the strip, unwind all twists save one, and join the cut up again. Because you removed an even number of twists, gluing together the cut edges rejoins points that started out near each other. This does not happen with an odd number of twists: one side of the cut is flipped end-to-end compared to the other.

For similar reasons, all bands made with an even number of twists are topologically the same as an ordinary cylindrical strip, which has no twists. However, the exact number of twists also has topological significance, because it affects how the band sits in its surrounding space. There are two different questions here, one about the intrinsic geometry of the band, the other about a band embedded in space. The first

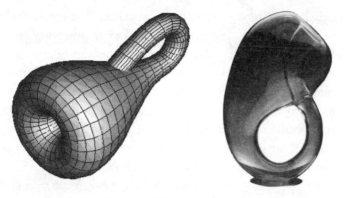

FIG 61 Klein bottle in (a) mathematics and (b) glass.

depends only on the parity (odd or even) of the number of twists; the second depends on the exact number.

The Möbius band has a boundary – those parts of the edge of the strip that don't get glued together. A sphere has no boundary. Can a one-sided surface have no boundary? It turns out that the answer is 'yes', a famous example being the Klein bottle (Figure 61). In this picture the 'spout' or 'neck' has been bent round, passed through the bottle's surface, and joined to the main bottle from the inside. In this representation, the Klein bottle meets itself in a small circular curve. The topologist ignores that intersection when thinking about an ideal Klein bottle, because it is an artefact that arises when the surrounding space is three-dimensional. No such surface can exist in three-dimensional space without crossing through itself. This is no problem to topologists, who can imagine surfaces in space of higher dimensions, or even in no surrounding space whatsoever, but it is an unavoidable obstacle for model-makers and glass-blowers.

Imagine trying to paint the Klein bottle. You start on the 'outside' of the large bulbous part, and work your way down the narrowing neck. When you cross the self-intersection, you have to pretend temporarily that it's not there, so you continue to follow the neck, which is now inside the bulb. As the neck opens up, to rejoin the bulb, you find that you are now painting the *inside* of the bulb! What appear to be the inside and outside of the Klein bottle connect together seamlessly: it is indeed one-sided.

Alan had heard that if you cut a Klein bottle along a suitable curve, then it falls apart into two Möbius bands, and he verified this in glass (Figure 62). If you do this with a Klein bottle that sits in ordinary space like Figure 61, those bands have a single twist. He wondered what sort of shape you had to cut up to get two three-twist Möbius bands. So he made a lot of different shapes in glass, cut them up, and saw what he

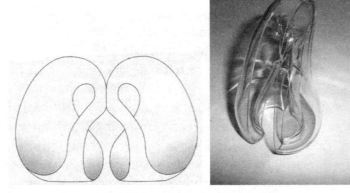

FIG 62 Cutting a Klein bottle into two Möbius bands (a) mathematically (b) in glass.

got. As he writes: 'I always like to solve problems in a practical way. I find that if enough variations to the basic concept are made, or collected, the most logical or obvious solution to the problem usually becomes apparent. In this case, adhering to as few limiting principles as possible, I started designing and making all sorts of single surface vessels. The basic Klein bottle can easily be stretched and distorted to form numerous shapes, but I wanted to go beyond this, and create new concepts. As far as I know my designs are all new; even so, they are traceable back to Klein's original bottle.'

Since he was looking for three-twist Möbius bands, Alan tried all sorts of variations on the number three – such as bottles with three necks (Figure 63), and, amazingly, sets of three bottles nested inside each other (Figure 64). He stacked three bottles on top of each other, and linked sets of three such stacks together. He thought about what would happen when they were cut up; he even cut them up with a diamond

FIG 63 Klein bottle with three necks.

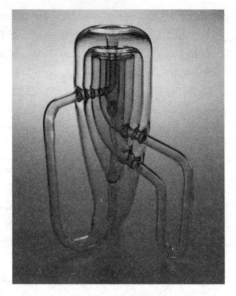

FIG 64 Nested set of three Klein bottles.

saw, to check. He started to 'see', in his mind's eye, the lines along which these shapes should be cut to make Möbius bands. But the three-twist bands proved elusive. The breakthrough was a very curious bottle whose neck looped round twice, forming three self-intersections (Figure 65). He named this the 'Ouslam Vessel' after the mythical bird that goes round in ever decreasing circles until it vanishes up its own rear end. 'Oozalum' is another common spelling.

If the Ouslam Vessel is sliced vertically, through its plane of left–right symmetry – the plane of the paper in the drawing – then it falls apart into two three-twist Möbius bands. Problem solved! But that was only the beginning. Like any mathematician, Alan was now after bigger game. What about five-twist

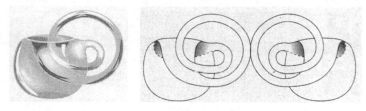

FIG 65 The 'Ouslam Vessel', here shown (a) in cross-section, whose neck loops round twice. If sliced as in (b) it falls apart into two three-twist Möbius bands.

bands? Seven-twist bands? Nineteen-twist bands! What was the general principle? Generalizing Figure 65 by adding in an extra loop, he quickly saw that five-twist bands would result. Every extra loop puts in two more twists.

Then he simplified the design, making it more robust, to produce spiral Klein bottles like Figure 66. This one cuts into two seven-twist bands – and every spiral turn you add puts in two more twists.

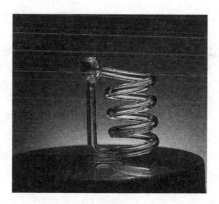

FIG 66 Spiral Klein bottle cuts into two seven-twist bands.

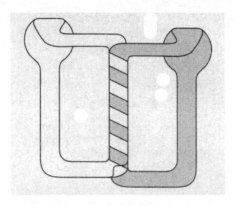

FIG 67 Cutting an ordinary Klein bottle along a spiral curve.

Having now seen the significance of spiral turns, Alan realized that he could go back to the original Klein bottle by 'untwisting' the spiral. The line along which the spiral Klein bottle should be cut would deform too. As the spiral neck of the bottle untwisted, the cut line twisted up. So, if you cut an ordinary Klein bottle along a spiral curve (Figure 67), then you can get as many twists as you want – in this case, nine.

Now for a final curiosity. The original motivation for the work was the possibility of cutting a Klein bottle to get two one-twist Möbius bands. But you can also cut a Klein bottle along a different curve, to get just *one* Möbius band. I'll leave you to work out how: answer below.

FEEDBACK

Robert L. Henrickson of Billings, Montana provided some fascinating information about similar bottles in pottery. *The Life, The Times, and the Art of Branson Graves Stevenson* by Herbert C. Anderson Jr (Jahner Publishing 1979) reports that 'In response to a challenge from his mathematician son, Maynard, Branson made his first Klein bottle using the topology suggestion of the German mathematician, Klein. He failed in his first try, until the famous English potter, Wedgewood, came to Branson in a dream and showed him how to make the Klein bottle. Branson followed Wedgewood's instructions and succeeded!' This was around fifty years ago. The book includes a picture of the pottery Klein bottle. It has a spout, which is not essential to the topology. Branson saw this as evidence for the power of the subconscious mind. His study of claywork and pottery led to the formation of the Archie Bray Foundation in Helena, Montana.

ANSWER

Figure 68 shows Bennett's method for cutting a Klein bottle along a different curve to get just *one* Möbius band.

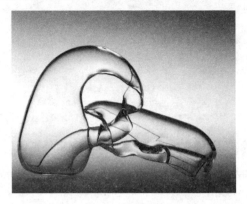

FIG 68 How Alan Bennett cuts a Klein bottle to form a single Möbius band.

WEBSITES

GENERAL:

http://en.wikipedia.org/wiki/Klein_bottle
http://plus.maths.org/issue26/features/mathart/index-gifd.html
http://mathworld.wolfram.com/KleinBottle.html
http://www.youtube.com/watch?v=E8rifKlq5hc

GLASS KLEIN BOTTLE:

http://www.kleinbottle.com/meter_tall_klein_bottle.html
http://www.kleinbottle.com/
http://www.sciencemuseum.org.uk/objects/mathematics/1996-545.aspx

17

Cementing Relationships

Art and science often seem poles apart, but every so often an artist manages to embody significant scientific ideas in a painting, dance, or sculpture. Jonathan Callan's curious cratered landscapes are based on the physical properties of cement. But mathematics is not far away.

THE PRESTIGIOUS SCIENTIFIC journal *Nature* – in which Francis Crick and James Watson published their epic discovery of the double-helix structure of DNA, for instance – manages to combine high-powered scientific research with a journalistic streak. For a time, one of its regular columns was Art and Science, written by art historian Martin Kemp. The column of 11 December 1997 described the remarkable landscapes of a London artist called Jonathan Callan. Conventional landscapes are paintings of natural scenery, but Callan's works are sculptures. They are landscapes with a difference, landscapes unlike anything seen on Earth. They are three-dimensional forms created by pouring cement onto a board drilled with a random set of holes (Figure 69).

Kemp, who is an Emeritus Research Professor in the History of Art Department at Oxford University, relates Callan's sculptures to work in complexity theory about sand-piles and 'self-organized criticality'. In a letter to the editor, Adrian Webster of the Royal Observatory, Edinburgh, pointed out that the curious geometry of Callan's landscapes

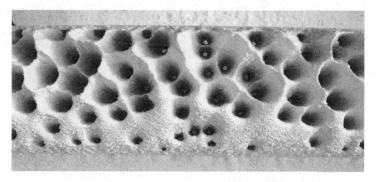

FIG 69 One of Jonathan Callan's landscapes.

can be understood using a much more classical branch of
mathematics, the theory of Voronoï cells. He also explained
how the Voronoï cells in Callan's landscapes illustrate one of
the big discoveries of recent astronomy, the foam-like distribution of matter in the Universe.

If ever there was an example of the unity of mathematics,
art, and science, this has to be it.

Kemp points out that artists have always relied on processes from physics and chemistry in their work – the frac-
ture of rock in classical sculpture, the properties of pigments,
even the flow of hot metal when casting bronze. However, the
traditional artist's technique has been to control these processes so that the media behave in desired ways. Callan is one
of a much smaller band of modern artists who allow the
physical and chemical processes of their media to determine
the main artistic features of their work – 'free-style evolution
of morphology', as Kemp puts it. The particular series of
works that excited the attention of *Nature* begins with surfaces
drilled with a random series of holes. The artist then sieves

cement powder evenly over the surface. Some cement trickles away through the holes, but further from the holes the powder piles up to form fantastic peaks, surrounding crater-like depressions centred on the holes.

Callan described the results like this: 'A de-natured geological principle, of sedimentary deposits, the silting of a river estuary…a geography that seems both eminently "natural" and highly "artificial" – the Alps brand new.' Kemp remarks that certain general principles seem to govern Callan's fantastic landscapes – for instance, the highest peaks occur in regions furthest removed from holes.

It is these regularities that Webster's work explains.

Civil engineers often have to work with soil – for example, their buildings usually rest on it. Roads that pass through cuttings in soft soil also require an understanding of how granular materials such as soil, sand, or cement pile up. The simplest and most important feature is the existence of a *critical angle*. Depending on the nature of the granular material, there is a steepest slope that it can sustain without collapsing. This slope runs at a constant angle, the critical angle. If you keep piling sand higher and higher, say by pouring it in a thin stream from a point above a fixed location, it will increase its angle of slope until it reaches this critical angle. Any extra soil will then trickle down the resulting pile, either causing a tiny avalanche or a big one, to restore the critical angle. The resulting 'steady-state' shape, in this simplest model, is a cone whose sides slope at exactly the critical angle (Figure 70a).

Complexity theorists study the process by which the slope attains this shape, and the nature of the avalanches, big or

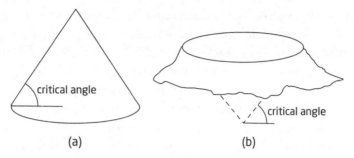

FIG 70 (a) Conical sand-pile. (b) Inverted conical crater.

small, that accompany its growth. The Danish physicist Per Bak coined the term 'self organized criticality' for such processes, and he has suggested that they model many important features of the natural world, especially evolution (where the avalanches involve not grains of soil, but entire species, and the piles are in an imaginary space of potential organisms). A real sand-pile is more complex than either the engineer's cone or Bak's avalanche, but it is useful as a metaphor.

Webster begins by noting that in Callan's art the structure of cement powder round a hole is complementary to the engineer's conical pile. Consider a horizontal board with just one hole. Away from the hole, cement rises in every direction at the critical angle, creating a conical depression whose tip points downwards and rests at the centre of the hole (Figure 70b). These inverted cones are the craters and canyons that form Callan's striking landscapes. In a simple model, they also slope at the critical angle.

But what of the geometry when there are several holes? The key point now is that any cascading cement powder that rolls down a slope and falls out through a hole will fall out through the hole that is *nearest* to its initial point of impact. This is a consequence of the slopes all being at the same angle. It is therefore possible to predict where the boundaries between the conical craters will occur. Divide the board into regions surrounding each hole, in such a manner that each region consists of precisely those points that are closer to the chosen hole than they are to any other hole. The hole's 'sphere of influence', so to speak – except that it is not a sphere, but a polygon. Provided the board is horizontal, the boundaries between these regions are directly beneath the common boundaries of adjacent craters.

Another way to describe these regions is to choose any pair of holes and draw a line between them, from centre to centre. Cut it in half, and from that point draw the line at right angles to it. That is, draw the perpendicular bisector of the two holes, considered as points located at their centres. Repeat this procedure for every pair of holes to get a network of lines. For each hole, find the smallest convex region that is bounded by segments of this network and contains that hole (Figure 71). This region is the *Voronoï cell* corresponding to the chosen hole. Each hole is surrounded by a unique Voronoï cell, and the Voronoï cells together tile the plane.

George Voronoï was a Russian mathematician who worked on number theory and multidimensional tilings around 1900, and his concept was taken up by the early crystallographers. Voronoï cells go by several other names – Dirichlet domains, Brillouin zones, and Wigner–Seitz cells – because they have been rediscovered independently in many contexts. The first person to define and study them in a technical sense seems to have been the mathematician Peter Lejeune-Dirichlet,

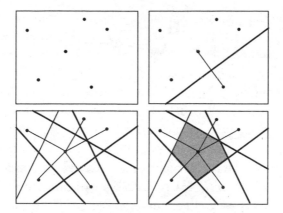

FIG 71 Construction of Voronoï cells.

who applied them to number theory in 1850, but René Descartes used them informally in 1644. In 1854 the British doctor John Snow used a Voronoï diagram in his famous study of cholera, to demonstrate that most of the victims lived nearer to the water pump in Broad Street than to any other pump – suggesting that the water from that pump was infected.

The geometry of Voronoï cells, combined with the critical angle for a sand-pile, imply that Callan's craters rise in inverted cones at the same critical angle. And they meet above the edges of the Voronoï cells defined by his system of drilled holes. One pleasant consequence of this geometry is that when two slopes meet, they come together along a well-defined ridge, with no sharp discontinuity. Another feature, less obvious, can also be deduced: the shape of these ridges, where one crater merges into its neighbour. In the abstract, two inverted cones rise at identical angles, so they must meet

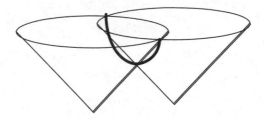

FIG 72 Callan's craters meet along hyperbolic ridges.

vertically above the perpendicular bisector of the line that joins their vertices. That is, the ridge lies directly above the Voronoï boundary. What curve do you get if you cut a cone with a vertical plane? The ancient Greeks knew the answer: a hyperbola (Figure 72). This fact helps to explain the rather jagged nature of Callan's landscapes, for where three Voronoï cells meet, we observe the intersection of three steeply rising hyperbolas.

What of the connection with galactic clusters? Astronomers have discovered that matter in the Universe is not uniformly spread out, but clumpy, forming loosely knit skeins surrounding huge voids (Figure 73). Theoretical models of this process involve Voronoï cells in three-dimensional space, with Callan's holes replaced by point masses. In the plane, the perpendicular bisector of a pair of points is a line, but in space it is a plane. Draw these bisecting planes for all pairs of points, and let the Voronoï cell of a given point be the smallest convex region that surrounds it and is bounded by portions of these planes. Now the Voronoï cell is a polyhedron. In the popular 'Voronoï foam' model of the Universe's distribution of matter, galaxies occur only on the boundaries between neighbouring Voronoï cells.

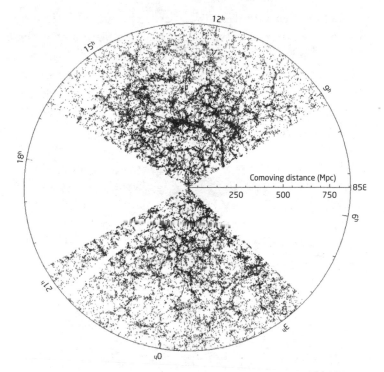

FIG 73 Distribution of galactic clusters, with large voids.

There is an analogy – loose rather than strict, but still illuminating – with the distribution of cement powder in Callan's landscapes. There, the cement piles up highest along the Voronoï boundaries. The analogous property in space would be that matter is *densest* along those boundaries. Because of the force of gravity, denser regions of matter draw nearby matter towards them, which concentrates matter more and more densely along the Voronoï boundaries. If Callan's cement exerted a gravitational force that could overcome frictional resistance between grains, its constituent grains would similarly migrate onto the polygonal foam-like network determined by the Voronoï boundaries. So this one simple idea encapsulates some arresting art, some elegant mathematics, and some deep physics about the distribution of matter in the Universe.

FEEDBACK

Originally I described Callan's landscapes as unlike anything found on 'any known world'. I've edited that remark out, because the landscape sculptures bear an uncanny resemblance to some of NASA's images of the surface of Hyperion (Figure 74). Hyperion is one of Saturn's many satellites, of which 61 have been detected as I write, with 53 confirmed and given official names. Is it possible that Hyperion is a dust-covered sponge, and the dust has slipped down cavities in the underlying rock? The satellite's low gravity would make the critical angle very steep, which seems to be consistent with the image.

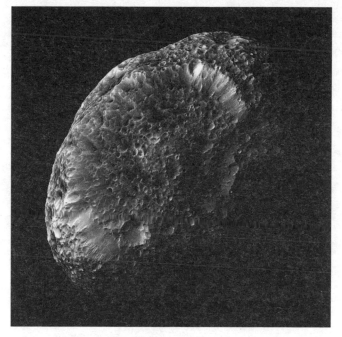

FIG 74 Hyperion (photo courtesy of NASA).

WEBSITES

VORONOÏ DIAGRAM:

  http://en.wikipedia.org/wiki/Voronoi_diagram
  http://mathworld.wolfram.com/VoronoiDiagram.html

JONATHAN CALLAN:

http://findarticles.com/p/articles/mi_m1248/is_3_89/
ai_71558227

CHOLERA EPIDEMIC:

http://en.wikipedia.org/wiki/John_Snow_(physician)

18

Knotting Ventured, Knotting Gained

The usual topological view of knots doesn't capture some of their more practical aspects, such as the thickness of the string or the existence of friction. Bearing these features in mind leads to the beginnings of a new theory, based on knotting real ropes together.

IN LESS THAN a century the mathematics of knots has gone from a minor curiosity to a major area of research that lies at the frontiers of the mathematical mainstream. Knots embody, in its purest form, one of the big problems in topology: to understand the different ways to position one geometrical form inside another. In the case of knots, the two forms are a circle, representing a closed loop of string, and the whole of three-dimensional space. As far as topologists are concerned, a knot is a circle that has been 'embedded in' three-dimensional space, in such a manner that it cannot be disentangled by continuously deforming the surrounding three-dimensional space.

This description is somewhat removed from everyday experience, where bits of string have *ends* and you deform the *string*, not the space. Nonetheless, it captures the 'knottiness' of knots rather well, as Colin C. Adams's *The Knot Book* shows. Certain practical aspects of knots, however, do not reduce so well to a topological formulation, and a clear case in point is the question of knotting two different lengths of string together. The main criterion here is that the join should not slip if you pull on the ends of the string. Surface friction and the material from which the string is made come into play, and the whole problem requires a different approach.

Nonetheless, there exists the beginnings of a mathematical theory, rather well suited for development by recreational mathematicians. It is the brainchild of Roger E. Miles at the Australian National University, Canberra, and is explained in his *Symmetric Bends*. 'Bend' was the word used by sailors for a method of knotting ropes together, back in the days when ships had sails and virtually everything on board was made from either wood or rope; it is still used by sailboat enthusiasts. Miles's main aim is to classify the geometry of bends in a systematic way, making it possible to search for new ones with desirable properties. The resistance of a given bend to slippage under tension can be determined experimentally by tying it and seeing what happens. The result offers a new slant on the mathematics of bits of string and the twiddly things you can make by wrapping them round themselves and each other.

The simplest and best-known bend is the reef (Figure 75a). In drawing such diagrams, of course, short breaks in the lines indicate which string passes over which; the strings themselves remain unbroken. One string is drawn with a light line, the other with a heavy one. Miles advocates the use of only horizontal and vertical lines, rather than sweeping curves, for several reasons: they are easier to draw, easier to understand, and they reveal the symmetry of the situation better (when there is any). Each string has one 'free' end – where it terminates – and one 'standing' end, shown by dotted lines, where it continues elsewhere. This diagram has

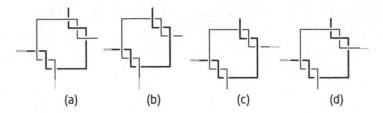

FIG 75 Four simple bends. (a) Reef. (b) Granny. (c) Whatknot.
(d) Thief knot. Note the difference between free ends (solid) and
open ends (dotted) which continue the string.

two types of crossing: dark-over-light or light-over-dark. In
more complex bends, there may also be dark-over-dark and
light-over-light crossings.

The reef, notoriously, is often confused with the granny
knot (Figure 75b). In traditional knot theory, where free ends
do not exist and everything is joined into loops, no other
knots are closely related to the reef and granny. Straight away
we discover that the situation is different for bends, for there
are two further bends which differ from reef and granny –
but only in the choice of which end is free. They are the
whatnot and the thief knot (Figures 75c, 75d).

These four 'elementary bends' are the ones with the
simplest diagrams, that is, the fewest crossings. Friction, pre-
venting strings from sliding, is to some extent generated at
crossings, and intuitively we would expect more complex
bends to be more secure – though not always, since security
also depends on how the sequence of crossings fits together
in three dimensions. All four elementary bends are highly
insecure, and tend to come apart if the strings are pulled or
otherwise disturbed. The way they come apart is instructive:

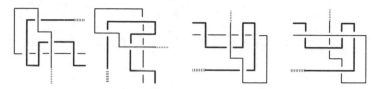

FIG 76 Three symmetry operations. (a) Original. (b) Diagonal flip. (c) 180° rotation. (d) Central inversion.

one string straightens out, though perhaps not completely, and then slides through the loops in the other string.

The elementary bends also have an appealing mathematical property: symmetry. The four bends just introduced exhibit three important symmetry operations (Figure 76). If the reef knot diagram is flipped over, keeping the diagonal from lower left to upper right fixed, then the same diagram appears – except that the colours (light/dark) are swapped. The same goes for the granny. The whatnot diagram looks the same, except for colour, if it is rotated through 180° about an axis pointing vertically out of the page. Finally the thief knot is symmetric under a 'central inversion' of three-dimensional space, which maps every point to the point on the same line through the origin, and the same distance away, but on the far side. That is, a point with coordinates (x, y, z) maps to $(-x, -y, -z)$. If you tie these bends with real string, and tighten them carefully and evenly, the resulting bends will possess the same symmetries.

There are more complicated bends too, of course. Indeed, Miles says that his interest in symmetric bends started in 1990, when he became aware of the 'rigger's bend' (Figure 77). The rigger's bend also has 180° rotational symmetry. It is

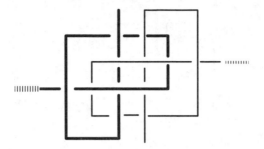

FIG 77 Rigger's bend.

often called 'Hunter's bend' after Dr. Edward Hunter, who discovered it in 1978. At the time it was thought to be new (it did not appear in the bible of the subject, the *Ashley Book of Knots*) but it can be found in the American mountaineer Phil Smith's *Knots for Mountaineering* of 1956. Miles first came across it in a copy of Mario Bigon and Guido Regazzoni's *The Morrow Guide to Knots* that he bought in San Francisco in 1989. By coincidence, it was on the San Francisco waterfront in 1943 that Smith invented the rigger's bend.

Based on the above three types of symmetry (diagonal flip, rotation, central inversion), Miles developed a formalism for studying and indeed inventing symmetric bends. An example of an entire family of bends found in this way is the generalized thief knot (Figure 78). However, there is more. There are three more symmetry operations (Figure 79) that can be performed on bends in three-dimensional space:

Mirror image: reflect the bend in a mirror. On a two-dimensional diagram, with the mirror in the plane of the page, this has the effect of reversing the crossings at every intersection.

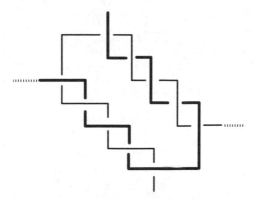

FIG 78 Generalized thief knot.

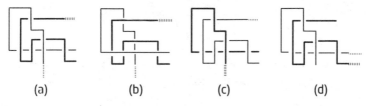

(a)   (b)   (c)   (d)

FIG 79 Three more symmetry operations. (a) Original. (b) Mirror image (in plane of paper). (c) Colour interchange. (d) Reverse.

Colour interchange: swap light and dark colours.
Reverse: interchange the dark standing and free ends, and at the same time interchange the light standing and free ends.

Any one of these operations changes a centro-symmetric bend to a centro-symmetric bend, and a bend with rotational symmetry to a bend with rotational symmetry.

The prize specimen in this context is the 'rewoven figure-of-eight bend', otherwise called the 'Flemish bend' (Figure 80).

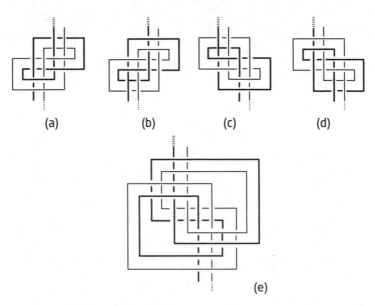

FIG 80 (a) Flemish bend. (b) Its mirror image. (c) Its reversal.
(d) Reversal of its mirror image. (e) The chameleon.

The first four diagrams show the Flemish bend, its mirror-image, its reversal, and the reversal of its mirror image. All four diagrams are rotationally symmetric. The fifth diagram has a different symmetry: it is centro-symmetric. Yet all five diagrams are topologically equivalent, that is, they can be continuously manipulated into each other! The easiest way to see this is to manipulate the fifth diagram into each of the others: I'll leave you the fun of finding out how. So a topological deformation can change the symmetry type of a bend, and Miles therefore renamed this bend the 'chameleon'.

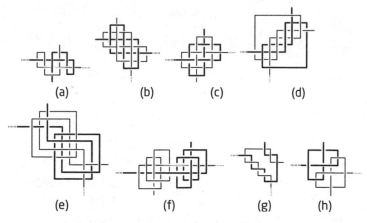

FIG 81 Eight bends. (a) Tight bend. (b) Tweedledee. (c) Crown bend. (d) Threefold. (e) Tweedledum. (f) Grapevine knot. (g) Surgeon's knot. (h) Pivotal knot.

Miles's book includes a catalogue of sixty symmetric bends, a selection of which is shown in Figure 81. Is there, he asks, a 'best' symmetric bend? His answer: 'not really.' The reason is that bends may be preferred for many different features. These include ease of tying, ease of checking that they have been tied correctly, ease of adjustment to make the free ends longer or shorter, tightness, resistance to jogging or tugging, compactness, streamlinedness, strength, ease of untying, beauty, charisma...

FEEDBACK

Mainstream mathematics has taken up the challenge of finding a more geometric theory of knots, though in a different way. The usual way to study knots topologically is in terms of invariants - properties that remain unchanged by deformations. Two knots that have different invariants must be topologically different. The first important invariant was the Alexander polynomial, discovered in the 1920s by J.W. Alexander. This is an algebraic expression associated with any knot, and knots whose Alexander polynomials are different cannot be deformed into each other. Unfortunately, knots with the same Alexander polynomial need not be topologically equivalent, the reef and granny knots being the simplest example. A newer topological invariant, the Jones polynomial, often succeeds where the Alexander polynomial fails; the Jones polynomial of a reef knot is different from that of a granny knot.

By making the 'string' of a knot more like physical string, mathematicians have discovered a new invariant, which is not a polynomial, but a number. The underlying idea goes back to I. Fary in 1929. Imagine tying a knot in a long rubber rod. The more complicated the knot, the more you have to bend the rod to tie it, so the more elastic energy the knotted rod acquires. Physical systems minimize energy, so we can ask what shape for the rubber rod makes the energy as small as possible.

In 1987 S. Fukuhara realized that there is a more convenient physical model: electrostatic energy. Think of the knot as a flexible wire of fixed length, which can pass through itself if necessary and is charged with static electricity. Because like charges repel each other, a knot that is free to move will arrange itself to keep neighbouring strands as far apart as possible, in order to minimize its electrostatic energy. This minimum energy value is the new geometric invariant. In 1991, Jun O'Hara of Tokyo Metropolitan University proved that the minimum energy of a knot increases as the knot becomes more complicated. Only a finite number of topologically different knots exist with energy less than or equal to any chosen value. This means that there is a natural numerical scale of complexity for knots, ranging from simple knots at the low-energy end to more complicated ones higher up.

What are the simplest knots? In 1993 a team of four topologists – Steve Bryson, Michael Freedman, Zhenghan Wang, and Zheng-Xu He – proved that the simplest 'knots' are exactly what you would expect. They are 'round circles', that is, circles in the everyday sense. Topologists, whose 'circles' are usually bent and twisted, have to append an adjective to remind themselves when they're not. In natural units, the energy of a round circle is 4, and all other closed loops have higher energy. Any loop with energy less than $6\pi + 4$ is topologically unknotted – it is a bent circle. More generally, a knot with c crossings in some two-dimensional picture has energy at least $2\pi c + 4$, though this bound is probably not the best possible, as the lowest-known energy for a trefoil knot – which has three crossings – is about 74, a lot bigger than $6\pi + 4 = 22.84$. The number of topologically distinct knots with energy less than or equal to E is at most 0.264×1.658^E.

WEBSITES

GENERAL:

http://en.wikipedia.org/wiki/Knot
http://www.animatedknots.com/
http://www.layhands.com/Knots/Knots_KnotsIndex.htm

RIGGER'S BEND:

http://en.wikipedia.org/wiki/Hunter%27s_bend

KNOT ENERGY:

http://en.wikipedia.org/wiki/Knot_energies
http://torus.math.uiuc.edu/jms/Videos/ke/

19

Most-Perfect Magic Squares

Combinatorics is the art of counting things without actually listing them – usually because the lists would be too big to fit inside the current Universe. One of the major open problems in recreational mathematics is to count the magic squares of a given size. For an important class of squares, we now know the answer.

'VE ALREADY MENTIONED magic squares several times, but let's recap. Take, say, the consecutive whole numbers from 1 to 16 and arrange them in a 4 × 4 array so that every row of four numbers, every column, and the two diagonals all add up to the same total. If you succeed – Figure 82 shows an example – you've made a magic square of order 4, and the common total is its 'magic constant'. Here the magic constant is 34, as it must be for all magic squares formed from the integers 1–16. If you do the same with the numbers from 1 to 25 in a 5 × 5 array, you've got a magic square of order 5, and so on. Magic squares are a favourite topic in recreational mathematics, and it is in the nature of favourite topics that they never become exhausted. Despite the vast literature on magic squares – and I *mean* vast – it always seems possible to put a new spin on the concept.

What is much harder, though, is to make a fundamental new contribution to the basic mathematics of the topic – one that goes beyond solely recreational interests and impinges upon the mathematical mainstream. Just such a contribution was published in 1998 by Dame Kathleen Ollerenshaw

1	15	14	4
12	6	7	9
8	10	11	5
13	3	2	16

FIG 82 A 4 × 4 magic square. All rows, columns, and diagonals sum to 34. Moreover, opposite pairs of numbers, relative to the centre, sum to 17.

and David S. Brée as *Most-Perfect Pandiagonal Magic Squares: Their Construction and Enumeration*.

In it they obtained the first significant partial solution of one of the big open problems in the subject: to count how many magic squares there are of any given order. Their main result is an explicit formula for the number of so-called 'most-perfect' squares of given order, together with systematic methods for constructing them all. In case this sounds like an easy problem, it is worth pointing out that the number of such squares of order 12 is more than 22 billion, while for order 36 it is roughly 2.7×10^{44}. You don't 'count' magic squares by writing them all out and chanting '1, 2, 3,...'.

Their work belongs to the area of mathematics known as combinatorics – the art of counting things without listing them. The result may have practical implications; indeed, the original stimulus came from potential applications of 8 × 8 magic squares to photographic reproduction and image processing.

A noteworthy feature of the research is its context, for neither author is a typical research mathematician. Dame Kathleen (the honour was awarded in 1971 for services to education) reached the age of 97 in October 2009, and spent most of her professional life in education and the upper reaches of university administration. Her collaborator David Brée has held positions in business studies, psychology, and most recently artificial intelligence.

For mathematical purposes it is more convenient to build a magic square of order n from the integers $0, 1, 2, \ldots, n^2{-}1$ rather than the traditional $1, 2, 3, \ldots, n^2$, and both the book and this chapter employ that convention. If you add 1 to every entry in a mathematician's magic square you get a traditional one, and conversely if you subtract 1 from every entry in a traditional magic square you get a mathematician's one. Thus there is no essential difference between the two conventions except for the square's magic constant, which is increased or diminished by n.

The magic constant of a traditional square of order n is $\frac{1}{2}n(n^2 + 1)$. That of a mathematical magic square of order n is $\frac{1}{2}n(n^2 - 1)$. There is a single magic square of order 1, namely

0.

There is no magic square of order 2 (the only order that never occurs) because the conditions force all four entries to be equal. There are eight magic squares of order 3, but they are all rotations or reflections of just one square

1	8	3
6	4	2
5	0	7

with magic constant 12. Obviously a rotation or a reflection of a magic square remains magic, so all magic squares of order 3 are 'essentially the same'. According to Chinese legend, the 'traditional' version of the above square (using the numbers 1–9, and known as the *lo-shu*) dates from around 2400 BC, where it was observed on the back of a turtle by the legendary Emperor Yu. Scholars consider this date to be questionable, and AD 1000 may well be more accurate.

There are 880 essentially different magic squares of order 4, an impressive 275,305,224 of order 5, and the number explodes as the order increases. No exact formula is known. By 'essentially different' I mean 'ignoring rotations and reflections'.

One way to make progress is to impose further conditions. For our purposes the most natural such condition is that the square should be *pandiagonal*, which means that all 'broken diagonals' must also sum to the square's magic constant. (Broken diagonals 'wrap round' from one edge to the opposite edge, Figure 83.) An example of a pandiagonal magic square is

0	11	6	13
14	5	8	3
9	2	15	4
7	12	1	10

with magic constant 30. Typical broken diagonals here are $11 + 8 + 4 + 7$ and $11 + 14 + 4 + 1$, both of which do indeed equal 30. There are 48 essentially different pandiagonal squares of order 4, and 3600 of order 5.

The order 3 square is not pandiagonal: for example $8 + 2 + 5 = 15$, not 12. More generally, Andrew H. Frost proved in 1878 that any even-order pandiagonal magic square must be 'doubly even', that is, a multiple of 4. A much slicker proof was given by C. Planck in 1919 – see Ollerenshaw and Breé's

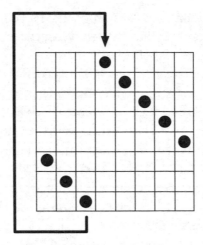

FIG 83 Broken diagonals.

book. Odd-order pandiagonal magic squares exist for all orders greater than 3.

Most-perfect squares, named by Emory McClintock in 1897, are even more restricted. As well as being magic and pandiagonal, they also have the property that any 2 × 2 block of adjacent entries give the same total, namely $2n^2-2$, where n is the order. Here we include 2 × 2 blocks that 'wrap round' from one edge to the opposite edge. It can be shown that any magic square with this property of 2 × 2 blocks is necessarily pandiagonal, but the converse is false.

The order 4 square above is most perfect – for example $0 + 11 + 14 + 5 = 30$, and $8 + 3 + 15 + 4 = 30$, and so on. An example of a 2 × 2 block that wraps round from one edge is the block 3, 4, 14, and 9.

More ambitiously, the order 12 square shown in Figure 84 is most perfect.

64	92	81	94	48	77	67	63	50	61	83	78
31	99	14	97	47	114	28	128	45	130	12	113
24	132	41	134	8	117	27	103	10	101	43	118
23	107	6	105	39	122	20	136	37	138	4	121
16	140	33	142	0	125	19	111	2	109	35	126
75	55	58	53	91	70	72	84	89	86	56	69
76	80	93	82	60	65	79	51	62	49	95	66
115	15	98	13	131	30	112	44	129	46	96	29
116	40	133	42	100	25	119	11	102	9	135	26
123	7	106	5	139	22	120	36	137	38	104	21
124	32	141	34	108	17	127	3	110	1	143	18
71	59	54	57	87	74	68	88	85	90	52	73

FIG 84 Order 12 most-perfect magic square.

The key to Ollerenshaw and Brée's counting method is a connection between most-perfect squares and 'reversible squares'. To explain what these are, we need some terminology. A sequence of integers has reverse similarity if, when the sequence is reversed and pairs of corresponding numbers are added, the totals are all the same. For example 1 4 2 7 5 8 has reverse similarity, because its reversal is 8 5 7 2 4 1 and the sums of corresponding numbers $1 + 8$, $4 + 5$, $2 + 7$, $7 + 2$, $5 + 4$, and $8 + 1$ are all equal – in this case to 9. A *reversible square* of order n is an $n \times n$ array formed by the integers 0, 1, 2,..., n^2-1 with the following properties:

- Every row has reverse similarity.
- Every column has reverse similarity.

- The sums of entries in the opposite corners of any rectangle are equal.

For instance, the array of integers in ascending order from left to right given by

0	1	2	3
4	5	6	7
8	9	10	11
12	13	14	15

is reversible. In the third row, for example, we have $8 + 11 = 9 + 10$, $10 + 9 = 11 + 8 = 19$, and the same kind of pattern holds for all other rows and all columns (though with totals other than 19). Moreover, equations such as $5 + 11 = 7 + 9$ and $1 + 15 = 3 + 13$ verify the third condition. A less trivial reversible square of order 12 is shown in Figure 85.

Reversible squares are generally not magic, as this example shows. However, Ollerenshaw and Brée show that every reversible square of doubly even order can be 'transformed' into a most-perfect magic square by a specific procedure, and every most-perfect magic square arises in this manner.

We illustrate the method on the above example. There are three steps:

1. Reverse the right-hand half of each row:

0	1	3	2
4	5	7	6
8	9	11	10
12	13	15	14

64	51	81	49	48	66	65	83	82	50	80	67
28	15	45	13	12	30	29	47	46	14	44	31
24	11	41	9	8	26	25	43	42	10	40	27
20	7	37	5	4	22	21	39	38	6	36	23
16	3	33	1	0	18	17	35	34	2	35	19
72	59	89	57	56	74	73	91	90	58	88	75
68	55	85	53	52	70	69	87	86	54	84	71
124	111	141	109	108	126	125	143	142	110	140	127
120	107	137	105	104	122	121	139	138	106	136	123
116	103	133	101	100	118	117	135	134	102	132	119
112	99	129	97	96	114	113	131	130	98	128	115
76	63	93	61	60	78	77	95	94	62	92	79

FIG 85 Order 12 reversible square.

2. Reverse the bottom half of each column

0	1	3	2
4	5	7	6
12	13	15	14
8	9	11	10

3. More complicated! For the order 4 case, it can be stated like this. Break the square up into 2 × 2 blocks. Move the four entries in each such block as shown in Figure 86. That is, the top left entry stays fixed, the top right moves diagonally two squares, bottom left moves two spaces to the right, and bottom right moves two spaces down. If anything falls off

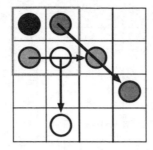

FIG 86 Transforming a reversible square to a magic one.

the edge of the 4 × 4 square, 'wrap the edges round' to find where it should go. For the general order n case, there is a similar recipe expressed by a mathematical formula. The result here is:

0	14	3	13
7	9	4	10
12	2	15	1
11	5	8	6

which you can check is indeed most perfect and magic.

There is a transformation process of this general type, setting up a one-to-one correspondence between most-perfect magic squares and reversible squares, for *any* doubly even order. Therefore you can count how many most-perfect magic squares there are, of a given doubly-even order, by instead counting how many reversible squares there are.

At first sight, this change in the nature of the problem doesn't get you very far, but it turns out that reversible squares have several nice features that makes it possible to count them systematically. In particular, reversible squares fall naturally into classes. Within each class, all members are related to each other by a variety of transformations, such as 'rotate', 'reflect', 'swap complementary pairs of rows', and a few more complicated manoeuvres. In order to construct all members of such a class, it is enough to construct just one of them and then routinely apply the transformations. Furthermore, each class contains precisely one special square, said to be 'principal', in which the top row starts with 0 1 and the integers in any row or column are in ascending order – so all you need to do is find *that* one.

Finally, each class has the same size. In fact, counting rotations and reflections of a given square as being 'essentially the same' and therefore not distinguishing between such squares, it can be proved that the number of essentially different squares in any class is

$$2^{n-2}([n/2]!)^2.$$

Here the exclamation mark indicates 'factorial', so that for example $6! = 6 \times 5 \times 4 \times 3 \times 2 \times 1 = 720$. It thus remains only to count the number of *principal reversible* squares of given order, and multiply that number by the formula just stated. The result will be the number of essentially different mostperfect magic squares of that order.

The number of principal reversible squares can itself be stated as a formula, though a rather complicated one. The discovery of this formula, and its proof, leads deeper into combinatorics, so I'll stop here, except to say that for doubly even orders $n = 4, 8, 12, 16$ the number of essentially different most-perfect magic squares is $48, 368{,}640, 2.22953 \times 10^{10}$, and 9.32243×10^{14}. The last two numbers are stated approximately here, but can be computed exactly. The number of essentially different most-perfect magic squares of order 144, incidentally, is 4.34616 × 10²⁵⁴, and again it is possible to write down all 255 digits if you really want to (computer assistance helps here).

FEEDBACK

Tom Hagedorn at the College of New Jersey sent me two articles about magic rectangles. A magic rectangle is an $m \times n$ array of square cells, filled with the integers from 1 to mn, arranged so that every row has the same sum and every column has the same sum. There is no requirement for the row sum to equal to column sum; in fact this is impossible if m and n are different. Moreover, diagonals are ignored. It has been known for more than a century that magic rectangles exist whenever m and n have the same parity (that is, they are both even or both odd), are bigger than 1, and are not both 2. Hagedorn generalizes this idea to higher dimensions, showing that if all the sides of the n-dimensional 'rectangle' are even, a magic rectangle exists.

The odd case is much harder. When I wrote the column in 1999 it was not known whether a $3 \times 5 \times 7$ magic rectangle exists. That is: can you put the numbers 1 to 105 into a

2	41	89	63	70
57	31	94	29	54
59	40	38	93	35
78	34	9	45	99
85	48	18	92	22
11	76	67	24	87
79	101	56	25	4

55	37	20	91	62
83	46	26	100	10
16	105	33	8	103
74	64	53	42	32
3	98	73	1	90
96	6	80	60	23
44	15	86	69	51

102	81	50	5	27
19	82	39	30	95
84	14	88	58	21
7	61	97	72	28
71	13	68	66	47
52	77	12	75	49
35	43	17	65	104

FIG 87 The 3 × 5 × 7 magic rectangle.

3 × 5 × 7 grid so that all horizontal rows have the same sum, all horizontal columns have the same sum, and all vertical columns have the same sum? These three sums may (must!) be different. The problem remained open until 2004 when Mitsutoshi Nakamura found such an arrangement (Figure 87).

WEBSITES

GENERAL:

http://en.wikipedia.org/wiki/Magic_square
http://mathworld.wolfram.com/MagicSquare.html
http://www.trump.de/magic-squares/

COUNTING:

http://www.maa.org/mathland/mathtrek_06_26_06.html

20

It Can't be Done!

Angle-trisectors and circle-squarers tend
to get annoyed when mathematicians
send back their work, saying that (a) it's
wrong, and (b) no, they haven't read it to
find the mistake. This is understandably
annoying. It is also perfectly fair and
entirely sensible. In mathematics, you
can prove a negative.

I N EVERYDAY LIFE, when we say something is impossible, we often don't mean that. Not literally, not absolutely. What we mean is that we can't see any way to achieve it. A lot of people thought that it was impossible for machines heavier than air to fly, and before that a lot of people thought that it was impossible for machines heavier than water to float – proving yet again that we never learn from history. Human ingenuity often overcomes apparent impossibilities. But even in everyday life, we can be confident that some things *are* impossible – human beings surviving unaided underwater for a year, say. (With suitable equipment, that's another matter.) And there's a grey area of things that most of us consider impossible but some believe in passionately, such as the ability to read another person's mind.

In mathematics, though, impossibility is something you can often *prove*. For instance, 3 is not an integer power of 2. One way to prove this is to ask what the power is, and observe that 2^1 is too small and anything from 2^2 upwards is too big.

It is true that the Bursar in Terry Pratchett's Discworld fantasy series believes that there is an extra whole number, 'umpt' – as in 'umpty-two' – but Roundworld mathematicians disagree. As this shows, impossibility proofs function only within the world of mathematics as it is currently set up: if you change the rules of the game, different things may happen. For example, in the world of integers 'modulo 5', in which any multiple of 5 is considered to be zero, then $3 = 2^3$. But that doesn't mean that my original impossibility statement is wrong, because the context has changed. It just means I have to be careful to define what I'm talking about. In textbook mathematics, that's very important, but in the Mathematical Recreations column I take a more relaxed approach, knowing that my readers (usually…) realize that I *could* be more precise if I wanted to.

This ability of mathematics to prove certain tasks impossible has a side-effect that can be frustrating. Imagine that I have spent the last ten years filling notebooks with long calculations, and I convince myself that I have discovered a new prime number, several thousand digits long. Unlike any other known prime, though, this one is *even*. Its final digit, in ordinary decimal notation, is 6. Excited beyond measure by this amazing feat, I send my work to a mathematician – who immediately sends it back again telling me that it's nonsense. Worse, when I ask him where I've made a mistake, he says that he hasn't read my work and he has no idea where the mistake is, but he knows that there must be one. I am appalled: what arrogance! I spent ten years on this problem; he spends ten minutes, ignores almost everything I've written, and yet claims he knows I'm wrong!

In most areas of everyday life, that would indeed be arrogance. But in mathematics, it is no more than a simple application of logic. The only even prime number is 2. There are no others. Why? Because even numbers are divisible by 2, and no prime exactly divides a *different* prime.

Kurt Gödel's proof that mathematics is undecidable – there is no algorithm for finding out whether a given statement has a valid proof – is one of the most profound impossibility theorems. Another big one comes from the nineteenth century, when Niels Henrik Abel, and later Évariste Galois, proved that the general equation of the fifth degree cannot be solved by a formula involving only the ordinary operations of algebra and the extraction of roots. Square roots, cube roots, fourth roots, whatever. Such expressions are called 'radicals'. Mathematicians of earlier ages had formulas in radicals for equations of the second, third, and fourth degree. Most of us learn the formula for the second degree ('quadratic' equations) at high school, which involves a square root; there are similar but increasingly complicated formulas for the third and fourth degree. All attempts to find a similar formula for the fifth degree failed.

Abel and Galois put a stop to such attempts by proving that they could never succeed. Abel's proof was a model of ingenuity; Galois's proof was more systematic, and required the creation of a new branch of mathematics, now known as Galois theory. Earlier, the Italian mathematician Paolo Ruffini had published a 500-page proof of the impossibility, and later published what he claimed was a simpler – though still gigantic – proof, but no one seemed convinced that there were no errors. Ironically, we now know that there was only

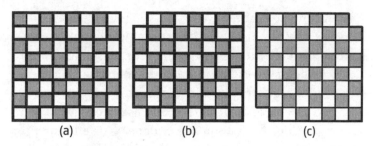

FIG 88 Boards (a) and (b) can be tiled with dominoes – for example, as shown. What about (c)?

one serious gap, and Abel filled it as part of his proof, without realizing that it completed Ruffini's.[6]

In order to see how such proofs are possible, consider a well-known puzzle. A chessboard has 64 squares. If you take 32 dominoes, each formed from two squares the same size as those of the chessboard, then there are enormously many ways to tile the board with dominoes – one is shown in Figure 88a. If you remove two adjacent corners of the board, you can easily tile the result with 31 dominoes – an example is Figure 88b. Throughout, I'm assuming as part of the conditions of the puzzle that each domino's squares coincide with two adjacent squares of the board, by the way. However, if you remove two *diagonally opposite* corners of the board (Figure 88c), all attempts to tile the result with dominoes fail.

Does your repeated failure *prove* that the task is impossible? No. Not even if you spent a lifetime trying. *Is* it impossible? Yes.

---

[6] For the history and context, see my book *Why Beauty is Truth*.

How can I be so sure?

Here's why. If you fit a domino on a chessboard, it always covers one black square and one white one. So, if you tile a board with dominoes, the number of white squares must be equal to the number of black squares. This is the case for the first two boards, but not for the one with opposite corners removed, which has 30 squares of one colour and 32 of the other.

This puzzle has a basic element in common with Galois's proof of the insolubility by radicals of the equation of the fifth degree. (Abel's proof does not fit quite so neatly into this framework.) Namely, the introduction of an *invariant*. This is some feature of a hypothetical solution that can be calculated without knowing the detailed form of that solution. For the domino problem, the invariant is a simple one: equality of black and white squares. For the fifth degree equation, it is a sophisticated algebraic feature of the symmetries of the roots of the equation, called the Galois group. If the invariant does not fit the conditions of the problem, whatever the proposed solution might be, then the proposed solution must fail. And you can tell that without even *seeing* the proposed solution!

If the invariant's wrong, your solution's wrong. That's it, there's no way out. It doesn't matter what your solution looks like.

Galois theory and recreational mathematics meet in a beautiful area of geometry: constructions using only an unmarked ruler and a compass.[7] A construction starts from

---

[7] Technically, the instrument concerned is a 'pair of compasses', but like a 'pair of scissors' this refers to a single gadget. A compass is a device that points north. But we must move with the times. And I was once asked why the constructions needed *two* compasses...

some known set of points, and successively locates new points as intersections of lines or circles. Any lines used must join known points, and any circles must be centred on a known point and pass through another known point.

What problems can be solved by such constructions? You can, for example, divide a given line segment into any specified number of equal pieces. You can divide a given angle into two equal angles (bisection), therefore also four equal parts, 8, 16,..., any power of two. You can draw regular polygons with 3, 4, 5, 6, 8, 10, and 12 sides. All this was known to Euclid. Over the next two millennia, many people tried to solve three other simple-looking problems by the same method:

- Duplicating the Cube: construct a cube whose volume is twice that of a given cube.
- Trisecting the Angle: trisect a given angle (cut it into three equal pieces).
- Squaring the Circle: construct a square whose area is equal to that of a given circle.

We now know why they had so much trouble: all three problems ask for the impossible.

We're not seeking approximate constructions here – it is straightforward to solve all three to any required degree of approximation. Nor are we asking for constructions that relax the conditions or employ other instruments. Figure 89 shows how to trisect an angle using either a marked ruler or a 'tomahawk'.

Again, the fact that nobody found an answer proves nothing. In 1796 Carl Friedrich Gauss discovered a ruler-and-compass construction for the regular 17-sided polygon which

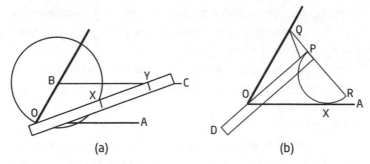

(a)   (b)

FIG 89 Trisecting angle AOB. (a) With marked ruler. Draw a circle
centre B through O. Draw BC parallel to OA. Mark X and Y on a ruler
with XY = OB. Slide the ruler until it passes through O, X is on the circle
and Y is on BC. Then angle AOY is one third of angle AOB. (b) Make a
tomahawk: a semicircle with diameter PR, extend PR to Q where PQ is
one half of PR, and PD perpendicular to PR. Arrange the tomahawk so
that PD passes through O, Q lies on OB, and OA is tangent to the semicircle (at X). Then angle POQ is one third of angle AOB.

had eluded all of his predecessors. Similar methods construct
regular polygons of 257 and 65,537 sides. Strange numbers –
why these? What else is possible? What isn't?

Specifically: what is the invariant for ruler-and-compass
constructions?

Any such construction can be represented in coordinate
form, and corresponds to the calculation of a sequence of
numbers, the coordinates of the points involved. Every step
in the construction turns out to introduce numbers that are
related to the known ones by an algebraic equation of degree
either 1 or 2 (1 for line-meets-line, 2 if a circle is involved).
This means (with some work) that the 'degree' of any point
in the construction – the lowest degree equation of which it

is a solution – must be a power of 2. This is the simplest invariant, and it is good enough to kill off all three problems listed above.

Duplicating the cube is equivalent to solving the equation $x^3-2 = 0$, which is of the third degree. Since 3 is not a power of 2, this is impossible.

Trisecting an angle is also equivalent to solving an equation of the third degree. (This follows from trigonometry and the equation $\cos 3x = 4 \cos^3 x - 3 \cos x$.) So this is also impossible.

Squaring the circle is equivalent to finding an equation, whose degree is a power of 2, that is satisfied by π. But (a difficult theorem proved by Ferdinand Lindemann in 1882) π does not satisfy an equation of *any* degree. (By the way, $x-\pi = 0$ doesn't count here. The coefficients must be related to the coordinates of the starting points.)

This, then, is how mathematicians know that it is a waste of time to try to solve any of these three problems using an unmarked ruler and compasses. If you want more details, see my textbook *Galois Theory*. Unfortunately, the existence of an impossibility proof does not stop people trying – probably because of a misunderstanding of the nature of mathematical impossibility. Underwood Dudley's fascinating book *A Budget of Trisections* records many such attempts.

The sad thing here is that trying to trisect the angle with ruler and compass is equivalent – via the invariant just described – to attempting to prove that 3 is an integer power of 2. Do you really want to go down in history as someone who thought they had proved *that*?

WEBSITES

TRISECTION:

http://en.wikipedia.org/wiki/Angle_trisection
http://mathworld.wolfram.com/AngleTrisection.html
http://www.cut-the-knot.org/pythagoras/archi.shtml

DUPLICATING THE CUBE:

http://en.wikipedia.org/wiki/Doubling_the_cube
http://www-history.mcs.st-and.ac.uk/HistTopics/Doubling_
  the_cube.html
http://mathforum.org/dr.math/faq/davies/cubedbl.htm

SQUARING THE CIRCLE:

http://en.wikipedia.org/wiki/Squaring_the_circle
http://en.wikipedia.org/wiki/Transcendental_number
http://mathworld.wolfram.com/CircleSquaring.html

QUINTIC EQUATION:

http://en.wikipedia.org/wiki/Quintic_equation
http://mathworld.wolfram.com/QuinticEquation.html

21

Dances with Dodecahedra

There are lots of ways to use mathematics, and to teach it. But here's an approach that had never occurred to me until its inventors told me about it. Unlike most mathematical recreations, this one is social. In fact, it sometimes needs ten people. Dancing.

I N CHAPTER **14** we took a new look at the ancient art of string figures – a topic that typically appeals to the mathematically minded, even though it isn't overtly mathematical. My confidence that the topic really *was* mathematical was to some extent justified by communications from readers, some of which are reported in the Feedback section of that chapter. One letter, however, raised a topic that was very different from anything I had anticipated: connections between string figures, mathematics, and dance. This was so interesting that it became a Mathematical Recreations column in its own right.

There are plenty of connections between mathematics and the arts – the use of perspective in painting and the ratios that occur in musical scales, for example – but the only link between mathematics and dance that I've seen before is an analysis of the symmetries of English country dancing carried out some years ago by my colleague Chris Budd, a mathematics professor at the University of Bath. The letter told me about something very different: the conscious use of mathematics to create new dances. It was from Karl Schaffer,

Co-Artistic Director of the Dr. Schaffer and Mr. Stern Dance Ensemble, Santa Cruz, and it described dances constructed around the use of several loops of string to create regular polyhedra and other mathematical figures.

Schaffer started by saying that he and Scott Kim had become interested in the topic of polyhedral string figures because in 1994 they had created a dance performance 'Through the Loop, In Search of the Perfect Square', which they performed in Bay Area K-8 schools. This is one of five mathematical dance shows produced by the company around that time, all of them bringing mathematical ideas to a youthful audience in a surprising and non-threatening context. Scott Kim, by the way, is a familiar name to long-term readers of the Mathematical Games column: Martin Gardner based a column on Kim's invention of calligraphic art in which the same arrangement of 'letters', read right way up or upside down, leads to different – often opposite – words.

The development of the show involved a local string figure enthusiast, Greg Keith, who taught them some traditional two-person string figure dances. They soon developed new ideas of their own, including three-dimensional string patterns based on polyhedra. In January 1998 they presented some of their work at Gardner Gathering III, a conference held in Atlanta in Gardner's honour.

As a simple example, Figure 90 shows how two dancers can produce a string tetrahedron (with two edges doubled) using a single loop of rope. The first dancer stands to the left, the second to the right, with the loop passing between them. Each holds the end of the loop in their right hand, while

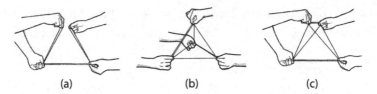

(a)   (b)   (c)

FIG 90 Two-person tetrahedron dance.

grasping both strands a little further along with their left hand. Simultaneously, dancer 1 crosses his right hand over his left, while dancer 2 separates her left and right hands. Then both reach forward with their right hands until they almost touch (Figure 90a). Next, each uses the right hand to grasp one strand of the other's rope, while continuing to hold on to their own portion of rope. Then dancer 1 slides his right hand along the double strand that it now holds, towards its natural position on his right, so that the rope looks like Figure 90b. Finally, both dancers raise their right hands and lower their left hands, and the result is a regular tetrahedron (Figure 90c) in which two sides are double strands and the other four are single strands.

In the same way, but with more left to your imagination, Figure 91 shows how six dancers, holding six loops of rope or ribbon, can produce the semi-regular polyhedron known as a 'cuboctahedron', which has six square faces and eight triangular faces. Figure 92 illustrates how the rope moves (but not the dancers!) for a more elaborate sequence. The dance begins with a single (long) loop, held by three people, which starts as a triangle and is manipulated first into a tetrahedron and then into an octahedron (a solid with eight triangular faces). Now a fourth dancer joins in, and helps to transform

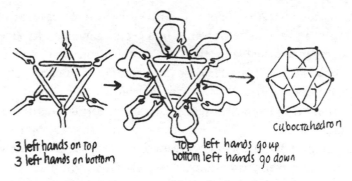

3 left hands on top
3 left hands on bottom

Top left hands go up
bottom left hands go down

cuboctahedron

FIG 91 Six-person cuboctahedron dance.

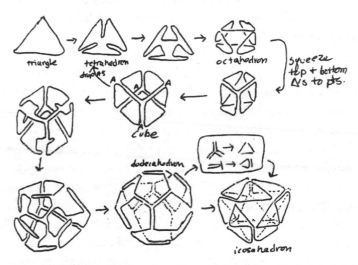

triangle

tetrahedron
doubles

octahedron

squeeze
top + bottom
△'s to pts.

cube

dodecahedron

icosahedron

FIG 92 Three/four/ten-person dance through all the regular polyhedra.

the octahedron into a cube. Finally, six more dancers join the dance, and the cube becomes first a dodecahedron (twelve pentagonal faces) and then an icosahedron (twenty triangular faces). All five Platonic solids (tetrahedron, cube, octahedron, dodecahedron, and icosahedron) are represented.

Schaffer remarks that sequences of transformations of this kind are easier to discover using actual strings than by making drawings on paper. Moreover, the search for new forms and transformations is necessarily a group activity, because you need enough hands to hold the strings. Usually each vertex of the polyhedron is held by only one hand, which is why ten people are need to form a dodecahedron with its twenty vertices. Arranging the participants so that the shape they are constructing can actually be seen by anybody else is decidedly tricky, though.

Experiments of this kind are good fun for a class of school students, and they provide a gentle introduction to three-dimensional thinking. At a deeper level, they can be used to develop serious mathematical ideas. For example, keeping track of which edges have to be doubled leads to a consideration of 'Euler cycles' in graphs. Recall that a graph is a collections of nodes linked by edges, and an Euler cycle is a closed path that passes along every edge. Here the nodes are the hands of the participants, and the edges are the edges of the polyhedron being made – realized physically by sections of rope. However, in the dances a single edge of a polyhedron sometimes corresponds to two or more strands of rope. Why? Can't it be done with only one strand per edge?

The answer, in general, is 'no'. Suppose for the sake of illustration that there is only one loop of rope. Then the rope forms a closed cycle that traverses every edge of the

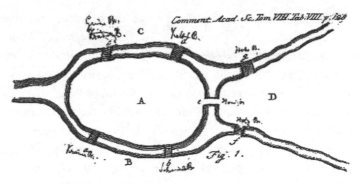

FIG 93 Euler's Königsberg bridges puzzle.

polyhedron. In 1735 Leonhard Euler encountered this question in connection with a famous puzzle: the Bridges of Königsberg. In the river Pregel, which flows through Königsberg, there are two islands. At that time, seven bridges linked the islands to the riverbank and to each other, as shown in Figure 93. The townspeople, so it is said, had spent many years trying to find a walking tour that passed over each bridge exactly once. Euler proved that no such path exists.

How? Euler's proof was symbolic, but can be interpreted as considering the four landmasses – two islands, two riverbanks – to be nodes, and the seven bridges to be edges, thereby turning the problem into a graph, or network. Then he proved that if such a cycle is required to pass along each edge of the graph exactly once, then an *even* number of edges must meet at every vertex. The key idea is that whenever the cycle encounters a node along one edge it must leave it along another, so the edges that meet that node fall into pairs – and so must be even in number. This condition fails for the Königsberg bridges, therefore no solution to the puzzle exists.

More significantly, Euler also proved the converse: for any connected (all in one piece) graph with the evenness property, a closed cycle passing along each edge exactly once always exists. Here his idea is to start by creating some closed cycle. If it happens to miss some edges, you can add these in by modifying the cycle so that it includes extra 'detours'. The evenness condition ensures that no detour ever gets 'stuck', unable to rejoin the original cycle. Continue adding detours until all edges have been included...Done!

This theorem lets us make sense of the doubled edges that turn up in the dances. Take the dodecahedron as an example. Here there are twenty nodes, the vertices, linked by thirty edges. Three edges (an odd number) meet at each vertex, so there cannot be a cycle in which each edge is traversed only once. However, if an edge is doubled up, then the vertex at each end is now met by four edges, which is even. Can you find ten edges which, when doubled up, produce an even number at *every* vertex? If not, you could double all edges: then six meet at every vertex. But do you really need that many? Incidentally, the dodecahedron in Figure 92 uses neither of these approaches, mostly because it involves three-fold rotational symmetry.

String loop dances can be used to illuminate many other areas of mathematics – simple ideas about three-dimensional geometry and symmetry, for instance. But your educational aims don't need to be as worthy as that: these dances are also enormous fun. In particular, they're great for breaking the ice at parties.

WEBSITES

DR. SCHAFFER AND MR. STERN DANCE ENSEMBLE:

http://www.mathdance.org

SCOTT KIM'S PALINDROMIC CALLIGRAPHY IS ANIMATED
AT:

http://www.scottkim.com/inversions

KÖNIGSBERG BRIDGES:

http://en.wikipedia.org/wiki/Seven_Bridges_of_K%C3%
B6nigsberg
http://mathworld.wolfram.com/KoenigsbergBridgeProblem.
html
http://www.contracosta.edu/math/konig.htm

GENERAL GRAPH THEORY:

http://en.wikipedia.org/wiki/Graph_theory

Further Reading

Chapter 1

Henry Ernest Dudeney, *Amusements in Mathematics*, Dover, New York 1958.

Ivar Ekeland, *The Broken Dice*, University of Chicago Press, Chicago 1993.

Martin Gardner, *Mathematical Magic Show*, Penguin, Harmondsworth 1965.

Ian Stewart, *Another Fine Math You've Got Me Into*, Freeman, New York 1992; reprinted Dover, New York 2003.

Ian Stewart, *Game, Set and Math*, Blackwell, Oxford 1989; reprinted Dover, New York 2007.

Ian Stewart, *How to Cut a Cake*, Oxford University Press, Oxford 2006.

Ian Stewart, *Math Hysteria*, Oxford University Press, Oxford 2004.

Chapter 2

Kenneth A. Brakke, The opaque cube problem, *American Mathematical Monthly* **99** (1992) 866–871.

Vance Faber, Jan Mycielski, and Paul Pedersen, On the shortest curve which meets all the lines which meet a circle, *Annales Polonici Mathematici* **154** (1984) 249–266.

Vance Faber and Jan Mycielski, The shortest curve that meets all the lines that meet a convex body, *American Mathematical Monthly* **93** (1986) 796–801.

Martin Gardner, The opaque cube problem, *Cubism for Fun* **23** (March 1990) 15.

Martin Gardner, The opaque cube again, *Cubism for Fun* **25** (December 1990) 14–15.

Bernd Kawohl, The opaque square and the opaque circle, in *General Inequalities VII*, International Series in Numerical Mathematics **123** (1997) 339–346.

Bernd Kawohl, Symmetry or not?, *Mathematical Intelligencer* **20** no. 2 (1998) 16–21.

Chapter 3

Cameron Browne, *Hex Strategy*, A.K. Peters, Natick MA 2000.

Martin Gardner, *Mathematical Puzzles and Diversions from Scientific American*, Bell, London 1961.

Sylvia Nasar, *A Beautiful Mind*, Faber & Faber, London 1998.

Ian Stewart, *Math Hysteria*, Oxford University Press 2004.

Chapter 4

Andrew Granville, Prime number patterns, *American Mathematical Monthly* **115** (2008) 279–296.

Harry L. Nelson, *Journal of Recreational Mathematics* **11** (1978–79) 231.

Andrew Odlyzko, Michael Rubinstein, and Marek Wolf, Jumping champions, *Experimental Mathematics* **8** no. 2 (1999) 107–118.

Chapter 5

A.H. Cohen, S. Rossignol, and S. Grillner (eds.), *Neural Control of Rhythmic Motions in Vertebrates*, Wiley, New York 1988.

P. Gambaryan, *How Mammals Run: Anatomical Adaptations*, Wiley, New York 1974.

M. Hildebrand, Symmetrical gaits of horses, *Science* **150** (1965) 701–708.

Eadweard Muybridge, *Animals in Motion*, Dover, New York 2000.

Chapter 6

Colin C. Adams, *The Knot Book*, W.H. Freeman, San Francisco 1994.

Colin C. Adams, Tilings of space by knotted tiles, *Mathematical Intelligencer* **17** no. 2 (1995) 41–51.

B. Grünbaum and G.C. Shephard, *Tilings and Patterns*, W.H. Freeman, New York 1987.

Chapter 7

Robert Geroch and Gary T. Horowitz, Global structure of space-times, in *General Relativity: An Einstein Centenary Survey* (editors S.W. Hawking and W. Israel), Cambridge University Press, Cambridge 1979, 212–293.

John Gribbin, *In Search of the Edge of Time*, Bantam Press, New York 1992.

H.G. Wells, *The Time Machine*, in *Selected Short Stories of H.G. Wells*, Penguin Books, Harmondsworth 1964.

Chapter 8

Jim Al-Khalili, *Black Holes, Wormholes and Time Machines*, Taylor and Francis, London 1999.

Jean-Pierre Luminet, *Black Holes*, Cambridge University Press, Cambridge 1992.

R. Penrose, Singularities and time-asymmetry, in *General Relativity: An Einstein Centenary Survey* (editors S.W. Hawking and

W. Israel), Cambridge University Press, Cambridge 1979, 581–638.

Edwin F. Taylor and John Archibald Wheeler, *Exploring Black Holes: An Introduction to General Relativity*, Addison-Wesley, New York 2000.

Chapter 9

Andreas Albrecht, Robert Brandenberger, and Neil Turok, Cosmic strings and cosmic structure, *New Scientist* 16 April 1987, 40–44.

Sean M. Carroll, Edward Farhi, and Alan H. Guth, An obstacle to building a time machine, *Physical Review Letters* **68** (1992) 263–269.

Marcus Chown, Time travel without the paradoxes, *New Scientist* 28 March 1992, 23.

John R. Cramer, Neutrinos, ripples, and time loops, *Analog* (February 1993) 107–111.

J. Richard Gott, III, Closed timelike curves produced by pairs of moving cosmic strings: exact solutions, *Physical Review Letters* **66** (1991) 1126–1129.

Michael S. Morris, Kip S. Thorne, and Ulvi Yurtsever, Wormholes, time machines, and the weak energy condition, *Physical Review Letters* **61** (1988) 1446–1449.

Ian Redmount, Wormholes, time travel, and quantum gravity, *New Scientist* 28 April 1990, 57–61.

Chapter 10

Donald G. Bancroft, *Rollable body*, US Patent #4,257,605, United States Patent and Trademark Office, Alexandria VA, 24 March 1981.

Alessandra Celletti and Ettore Perozzi, *Celestial Mechanics: The Waltz of the Planets*, Springer, New York 2006.

Richard S. Westfall, *Never at Rest: A Biography of Isaac Newton*, Cambridge University Press, Cambridge 1983.

Michael White, *Isaac Newton: The Last Sorcerer*, Fourth Estate, London 1998.

Chapter 11

J. Eggers and T.F. Dupont, Drop formation in a one-dimensional approximation of the Navier–Stokes equation, *Journal of Fluid Mechanics* 262 (1994) 205.

D.H. Peregrine, G. Shoker, and A. Symon, The bifurcation of liquid bridges, *Journal of Fluid Mechanics* 212 (1990) 25–39.

X.D. Shi, Michael P. Brenner, and Sidney R. Nagel, A cascade structure in a drop falling from a faucet, *Science* 265 (1994) 219–222.

D'Arcy W. Thompson, *On Growth and Form*, Cambridge University Press, Cambridge 1942.

Chapter 12

R.A.J. Matthews, The interrogator's fallacy, *Bulletin of the Institute of Mathematics and its Applications* 31 (1994) 3–5.

Chapter 13

Robert Abbott, *Supermazes*, Prima Publishing, Rocklin 1997.

Martin Gardner, *The Colossal Book of Mathematics*, W.W. Norton, New York 2001.

Martin Gardner, *More Mathematical Puzzles and Diversions from Scientific American*, Bell, London 1963.

Ian Stewart, A partly true story, *Scientific American* **268** no. 2 (1993) 85–87.

Chapter 14

W.W. Rouse Ball and H.S.M. Coxeter, *Mathematical Recreations and Essays*, Macmillan, London 1939.

Henry Ernest Dudeney, *Amusements in Mathematics*, Dover, New York 1958.

Maurice Kraitchik, *Mathematical Recreations* (2nd edn), Allen & Unwin, London 1960.

Allen J. Schwenk, Which rectangular chessboards have a knight's tour?, *Mathematics Magazine* **64** no. 5 (1991) 325–332.

Chapter 15

Joseph D'Antoni, Variations on Nauru Island figures, *Bulletin of the International String Figure Association* **1** (1994) 27–68.

Caroline Jayne, *String Figures and How to Make Them*, Dover, New York 2003.

James R. Murphy, Using string figures to teach math skills, *Bulletin of the International String Figure Association* **4** (1997) 56–74.

Mark A. Sherman, Evolution of the Easter Island string figure repertoire, *Bulletin of String Figures Association* **19** (1993) 19–87.

Yukio Shishido, The reconstruction of the remaining unsolved Nauruan string figures, *Bulletin of the International String Figure Association* **3** (1996) 108–130.

Alexei Sossinsky, *Knots*, Harvard University Press, Cambridge MA 2002.

Tom Storer, *Bulletin of String Figures Association* special issue **16** (1988) (especially Chapter III on Indian diamonds).

Kurt Vonnegut, *Cat's Cradle* (new edn), Penguin Books, Harmondsworth 1999.

Chapter 16

Stephan C. Carlson, *Topology of Surfaces, Knots and Manifolds: A First Undergraduate Course*, Wiley, New York 2001.

John Fauvel, Raymond Flood, and Robin Wilson (eds.), *Möbius and His Band: Mathematics and Astronomy in Nineteenth-Century Germany*, Oxford University Press, Oxford 1993.

Chapter 17

Martin Kemp, Callan's canyons: art and science, *Nature* **390** (11 December 1997) 565.

Adrian Webster, Letter to the editor, *Nature* **391** (29 January 1998) 431.

Chapter 18

Colin Adams, *The Knot Book*, W.H. Freeman, New York 1994.

Clifford W. Ashley, *The Ashley Book of Knots*, Faber & Faber, London 1993.

M. Bigon and G. Regazzoni, *The Morrow Guide to Knots*, Morrow, New York 1982.

Roger E. Miles, *Symmetric Bends*, World Scientific, Singapore 1995.

Phil D. Smith, *Knots for Mountaineering* (3rd edn), Citrograph, Redlands 1975.

Alexei Sossinsky, *Knots*, Harvard University Press, Cambridge MA 2002.

Chapter 19

W.S. Andrews, *Magic Squares and Cubes*, Dover, New York 2000.

Kathleen Ollerenshaw, *To Talk of Many Things*, Manchester University Press, Manchester 2004.

Kathleen Ollerenshaw and David S. Brée, *Most-Perfect Pandiagonal Magic Squares: Their Construction and Enumeration*, Institute of Mathematics and Its Applications, Southend-on-Sea 1998.

Frank J. Swetz, Legacy of the Luosho, A.K. Peters, Wellesley MA 2008.

Chapter 20

Underwood Dudley, *A Budget of Trisections*, Springer, New York 1987.

Underwood Dudley, *Mathematical Cranks*, Mathematical Association of America, Washington DC 1996.

Underwood Dudley, *The Trisectors*, Mathematical Association of America, Washington DC 1996.

Mario Livio, *The Equation That Couldn't Be Solved*, Souvenir Press, London 2006.

Ian Stewart, *Galois Theory*, CRC Press, Boca Raton 2003.

Ian Stewart, *Why Beauty is Truth*, Basic Books, New York 2007.

Chapter 21

Martin Gardner, *The Colossal Book of Mathematics*, W.W. Norton, New York 2001.

Robin J. Wilson, *Introduction to Graph Theory*, Longman, Harlow 1985.

Index